TISSUE BARRIERS IN DISEASE, INJURY AND REGENERATION

TISSUE BARRIERS IN DISEASE, INJURY AND REGENERATION

Edited by

NIKOLAI V. GORBUNOV

Henry M. Jackson Foundation for the Advancement of Military Medicine, 6720A Rockledge Drive
Bethesda, Maryland 20817
United States

ELSEVIER

Elsevier
Radarweg 29, PO Box 211, 1000 AE Amsterdam, Netherlands
The Boulevard, Langford Lane, Kidlington, Oxford OX5 1GB, United Kingdom
50 Hampshire Street, 5th Floor, Cambridge, MA 02139, United States

Library of Congress Cataloging-in-Publication Data
A catalog record for this book is available from the Library of Congress

British Library Cataloguing-in-Publication Data
A catalogue record for this book is available from the British Library

ISBN: 978-0-12-818561-2

For information on all Elsevier publications visit our website at
https://www.elsevier.com/books-and-journals

Publisher: Andre Gerhard Wolff
Acquisitions Editor: Glyn Jones
Editorial Project Manager: Naomi Robertson
Production Project Manager: Paul Prasad Chandramohan
Cover Designer: Matthew Limbert
Cover Image Credit: Julian M. Schneider, Ulm University Hospital, Germany

Typeset by TNQ Technologies

Contents

Contributors *ix*
Preface *xiii*

1. Roles and distribution of telocytes in tissue organization in health and disease **1**

Carmen Elena Condrat, Mădălina Gabriela Barbu, Dana Claudia Thompson, Cezara Alina Dănilă, Andreea Elena Boboc, Nicolae Suciu, Dragoş Creţoiu *and* Silviu Cristian Voinea

1. Introduction 1
2. Telocytes—definition and history 2
3. Localization and identification methods 3
4. Physiological functions of telocytes 7
5. Roles of telocytes in disease 19
6. Perspectives and conclusion 29

References 30

2. Effects of radiation on endothelial barrier and vascular integrity **43**

Roxane M. Bouten, Erik F. Young, Reed Selwyn, Diego Iacono, W. Bradley Rittase *and* Regina M. Day

1. Introduction: radiation-induced permeability of the vasculature 43
2. Direct mechanisms of radiation-induced vascular effects 45
3. Indirect mechanisms of radiation-induced vascular effects 70
4. Conclusions 79

Acknowledgments 79
References 80

3. How severe RNA virus infections such as SARS-CoV-2 disrupt tissue and organ barriers—Reconstitution by mesenchymal stem cell-derived exosomes **95**

Jason Sanders *and* E. Marion Schneider

1. COVID-19 and the contribution by immune effector cells 95
2. Immune-mediated pathogenesis 96

3. Acute lung injury, increased endothelial permeability, and loss of organ barrier function 97
4. Endogenous repair systems 97
5. The role of mesenchymal stem cells 99
6. Extracellular vesicles: Exosomes and small microvesicles 99
7. Tissue reconstitutive mechanisms by mesenchymal stem cell-small extracellular vesicles in COVID-19 100
8. Source of exosomes 103
9. Mesenchymal stem cell-small extracellular vesicles as investigational new drug 105
10. Exosome enrichment 105
References 107

4. The retinal pigment epithelium: at the forefront of the blood-retinal barrier in physiology and disease 115

Richárd Nagymihály, Yaroslav Nemesh, Taras Ardan, Jan Motlik, Jon Roger Eidet, Morten C. Moe, Linda Hildegard Bergersen, Lyubomyr Lytvynchuk *and* Goran Petrovski

1. Introduction 115
2. Inner blood-retinal barrier 117
3. Outer blood-retinal barrier 122
4. Structural changes on optical coherence tomography and fluorescein angiography in the internal blood-retinal barrier and outer blood-retinal barrier in disease states 134
5. Conclusions 136
Acknowledgments 136
References 136

5. Appressorium morphogenesis and penetration in rice blast fungus 147

Huan-Bin Shi, Xue-Ming Zhu, Yun-Ran Zhang, Xiao-Hong Liu *and* Fu-Cheng Lin

1. Introduction 147
2. Appressorium induction and differentiation 148
3. Appressorium maturation and function 150

4. Appressorium penetration and invasive growth 152
5. Conclusion 154
Acknowledgments 154
References 154

6. Mesenchymal stem cells and exosomes in tissue regeneration and remodeling: characterization and therapy 159

Juliann G. Kiang

1. Introduction 159
2. Mesenchymal stem cell characterization 160
3. Mesenchymal stem cells and tissue or organ therapy 166
4. Replacement of mesenchymal stem cells with exosomes including anti-inflammatory cytokines, growth factors, and micro-RNAs 175
5. Conclusions 177
Acknowledgments 177
References 178
Further reading 185

Index 187

Contributors

Taras Ardan
Laboratory of Cell Regeneration and Plasticity, Institute of Animal Physiology and Genetics, Czech Academy of Science, Libechov, Czech Republic

Mădălina Gabriela Barbu
Alessandrescu-Rusescu National Institute for Mother and Child Health, Fetal Medicine Excellence Research Center, Bucharest, Romania

Linda Hildegard Bergersen
Institute for Oral Biology, Faculty of Dentistry, University of Oslo, Oslo, Norway

Andreea Elena Boboc
Alessandrescu-Rusescu National Institute for Mother and Child Health, Fetal Medicine Excellence Research Center, Bucharest, Romania

Roxane M. Bouten
Department of Pharmacology and Molecular Therapeutics, Uniformed Services University of the Health Sciences, Bethesda, MD, United States

Carmen Elena Condrat
Alessandrescu-Rusescu National Institute for Mother and Child Health, Fetal Medicine Excellence Research Center, Bucharest, Romania

Dragoş Creţoiu
Alessandrescu-Rusescu National Institute for Mother and Child Health, Fetal Medicine Excellence Research Center, Bucharest, Romania; Department of Cell and Molecular Biology and Histology, Carol Davila University of Medicine and Pharmacy, Bucharest, Romania

Regina M. Day
Department of Pharmacology and Molecular Therapeutics, Uniformed Services University of the Health Sciences, Bethesda, MD, United States

Cezara Alina Dănilă
Alessandrescu-Rusescu National Institute for Mother and Child Health, Fetal Medicine Excellence Research Center, Bucharest, Romania

Jon Roger Eidet
Center for Eye Research, Department of Ophthalmology, Oslo University Hospital, Oslo, Norway

Diego Iacono
Departments of Neurology, Pathology, and Anatomy, Physiology and Genetics, Uniformed Services University of the Health Sciences, Bethesda, MD, United States; Neurodegenerative Clinic, National Institute of Neurological Disorders and Stroke (NINDS), NIH, Bethesda, MD, United States; Department of Anatomy, Physiology, and Genetics, Uniformed Services University of the Health Sciences, Bethesda, MD, United States

Juliann G. Kiang
Scientific Research Department, Armed Forces Radiobiology Research Institute, Uniformed Services University of the Health Sciences, Bethesda, MD, United States; Department of Pharmacology and Molecular Therapeutics, Uniformed Services University of the Health Sciences, Bethesda, MD, United States; Department of Medicine, Uniformed Services University of the Health Sciences, Bethesda, MD, United States

Fu-Cheng Lin
State Key Laboratory for Managing Biotic and Chemical Treats to the Quality and Safety of Agro-products, Institute of Plant Protection and Microbiology, Zhejiang Academy of Agricultural Sciences, Hangzhou, Zhejiang, China; State Key Laboratory of Rice Biology, Institute of Biotechnology, Zhejiang University, Hangzhou, Zhejiang, China

Xiao-Hong Liu
State Key Laboratory of Rice Biology, Institute of Biotechnology, Zhejiang University, Hangzhou, Zhejiang, China

Lyubomyr Lytvynchuk
Department of Ophthalmology, Justus-Liebig-University Giessen, Eye Clinic, University Hospital Giessen and Marburg GmbH, Giessen, Germany

Morten C. Moe
Center for Eye Research, Department of Ophthalmology, Oslo University Hospital, Oslo, Norway; Institute of Clinical Medicine, Faculty of Medicine, University of Oslo, Oslo, Norway

Jan Motlik
Laboratory of Cell Regeneration and Plasticity, Institute of Animal Physiology and Genetics, Czech Academy of Science, Libechov, Czech Republic

Richárd Nagymihály
Center for Eye Research, Department of Ophthalmology, Oslo University Hospital, Oslo, Norway

Yaroslav Nemesh
Laboratory of Cell Regeneration and Plasticity, Institute of Animal Physiology and Genetics, Czech Academy of Science, Libechov, Czech Republic

Goran Petrovski
Center for Eye Research, Department of Ophthalmology, Oslo University Hospital, Oslo, Norway; Institute of Clinical Medicine, Faculty of Medicine, University of Oslo, Oslo, Norway

W. Bradley Rittase
Department of Pharmacology and Molecular Therapeutics, Uniformed Services University of the Health Sciences, Bethesda, MD, United States

Jason Sanders, MD
Chief Medical Officer, EV Biologics, Inc., Wyoming, United States

E. Marion Schneider
Division of Experimental Anaesthesiology, University Hospital Ulm, Ulm, Germany

Reed Selwyn
Department of Radiology, University of New Mexico, Albuquerque, NM, United States

Huan-Bin Shi
State Key Laboratory for Managing Biotic and Chemical Treats to the Quality and Safety of Agro-products, Institute of Plant Protection and Microbiology, Zhejiang Academy of Agricultural Sciences, Hangzhou, Zhejiang, China; State Key Laboratory of Rice Biology, China National Rice Research Institute, Hangzhou, Zhejiang, China

Nicolae Suciu
Division of Obstetrics, Gynecology and Neonatology, Carol Davila University of Medicine and Pharmacy, Bucharest, Romania; Department of Obstetrics and Gynecology, Polizu Clinical Hospital, Alessandrescu-Rusescu National Institute for Mother and Child Health, Bucharest, Romania

Dana Claudia Thompson
Alessandrescu-Rusescu National Institute for Mother and Child Health, Fetal Medicine Excellence Research Center, Bucharest, Romania

Silviu Cristian Voinea
Department of Surgical Oncology, Prof. Dr. Alexandru Trestioreanu Oncology Institute, Carol Davila University of Medicine and Pharmacy, Bucharest, Romania

Erik F. Young
Department of Electrical Engineering, Columbia University, New York, NY, United States

Yun-Ran Zhang
State Key Laboratory of Rice Biology, Institute of Biotechnology, Zhejiang University, Hangzhou, Zhejiang, China

Xue-Ming Zhu
State Key Laboratory for Managing Biotic and Chemical Treats to the Quality and Safety of Agro-products, Institute of Plant Protection and Microbiology, Zhejiang Academy of Agricultural Sciences, Hangzhou, Zhejiang, China

Preface

SCOPE AND OBJECTIVE

This book is a mutual effort of a group of authors with expertise in translational research and biotechnology. The book chapters discuss the concept of "tissue barriers" in conjunction with pathogenesis of diseases related to trauma, injury, and infection. To the subject, the authors present their view on structural and functional organization of barrier-forming tissues and the barrier-to-pathogen interactions as well as elucidate the emerging role of barrier-forming tissues in the regenerative medicine. On this account, the readers who are working on the development of new remedies for therapy of stress, injury, and degenerative diseases could avail themselves of in-depth understanding of the biology of tissue barriers, the mechanisms of biogenesis, and remodeling of tissue barriers and regulation of the barrier function under pathological conditions. Giving a wide-angle perspective on biomedical aspects of tissue barriers, this book is addressed to a broad audience of readers from students to practicing clinicians and experts in tissue barrier research.

The role of barrier-forming components and barrier functions in health and disease

From the research paradigm epistemology, a singularity of biological barrier represents a dividing surface *or medium* formed by biomolecules and cellular structures that separate an individual organism, i.e., "*living matter,*" from surrounding "*nonliving matter*" and other organisms; and/or segregate systems and constituent parts within a living organism. Thus, the network of these barrier "bio-interfaces" provides compartmentalization of the organismal and cellular components, maintains their high degree of structural (i.e., low entropy) and functional organization, and controls flow of energy and "information" in the form of biological cues under thermodynamically nonequilibrated conditions. In conjunction with the above, tissue barriers represent structural/functional entities formed by specialized, immunocompetent, or sentinel type of cells, which outline internal organs, tissues, tracts, and the entire body of organisms, are ubiquitously present in the parenchymal tissues and which sustain organismal immunochemical, metabolic, and thermal homeostasis. From an ontogeny perspective, tissue barriers are evolutionary evolved and adapted to: (i) maintain organismal morphogenesis, integrity, and endurance; (ii) regulate numerous vital physiological processes and thermal balance of organisms; as well as (iii) proactively and reactively respond to environmental impacts and damage; (iv) mediate interactions with other organisms; and (v) repel virus invasions.

The tissue barriers are configured and executed by ensembles of the barrier-forming cells of different histogenetic origins along with their extracellular matrices and intercellular junctions. In animals, these cells are, namely: mesenchymal, integument adipose, epithelial, endothelial, perivascular, immune, and ubiquitously distributed interstitial telocytes and reticuloendothelial cells. And in plants, the tissue barriers are constituted by epidermal, vascular, and sclerenchyma cells. It is broadly accepted that the structural-functional architecture of the barrier-forming networks is based on complex molecular and cellular mechanisms, including communications via cell-cell contacts, inter alia exchange with cell constituents, phagocytosis; and interactions of the above. Moreover, each of these cell the networks is subjected to modifications due flow of to the intrinsic morphogenetic stimuli, external forces, and impacts of diverse homeostatic cues (e.g., stress, danger) originated from tissue parenchyma and interstitial and circulating fluids. To this matter, the book chapters discuss several crucial aspects of the emerging signaling hubs mediated by telocytes and mesenchymal stromal cells in the histogenesis and integration of the barrier-forming tissues.

It ought to be noted that while the barrier-forming tissue systems are predominantly constituted by nonproliferating (e.g., terminally differentiated and G0-stage) cells, evidently, they are capable of self-rejuvenation and replacement of aberrant cells by both regulation of the tissue-specific progenitors and by recruitment of adult stem cells of mesenchymal and endothelial origins from milieux of stromal and lymphoid tissues and vascular niches. These pathways represent the basic cellular mechanisms, which mediate reconstitution and remodeling of the barrier integrity. Furthermore, these recruited cells sustain numerous "housekeeping functions" such as replenishing of impaired cells as well as phagocytosis of dead cells and their debris. In this light, it is worth to emphasize that a major population of nonprofessional phagocytic cells in stroma and vasculature is represented by endothelial and stromal fibroblastic cells. Therefore, it is hard to overstate the role of these cell lineages and their precursors in the responses to trauma and injury and in tissue regeneration. In this respect, the book chapters have also discussed the cellular mechanisms implicated in barrier functions mediated by endothelial and mesenchymal cells.

A remarkable plasticity attributed to the tissue barriers allows them to form numerous dynamic lines of defense to resist infections despite the pathogen's "endeavor" to subvert the host countermeasure mechanisms. This "status quo" is deranged in case of barrier breakdown due to an open wound or when acute or chronic diseases lead to impairment of barrier functions. The latter occur subsequently after ionizing irradiation that can damage the barrier-forming cells. Thus, exposure to gamma- or X-rays can trigger acute radiation sickness due to impairment and death of radiosensitive barrier-forming cells, namely, hematopoietic, lymphoid, mucosal epithelial, and vascular endothelial. Moreover, ionizing irradiation causes decline of the cell reproductive and regenerative capacity due to radiation-induced suppression of cell clonogenic potentials. Thus, the impact of irradiation and subsequent reactive responses can ultimately lead to delayed attrition of immune, intestinal, and vascular barriers accompanied by fluid loss, interstitial edema, and bacterial sepsis culminating in the failure of organs and systems. Advanced management of the radiation morbidity and mortality considers administration of diverse remedies. That includes bone marrow and stem cell transplant and application of related cell growth factors to restore the barrier function. In conjunction, recent developments in regenerative medicine and tissue engineering based on exploration of mesenchymal stromal cells brought a new perspective toward cell therapy of radiation injury. Indeed, firstly, populations of mesenchymal stromal cells represent a source of the adult stem cells [i.e., colony-forming unit fibroblasts, (CFU-F)] able to self-renew as well as to differentiate into multiple barrier-forming cell lineages. Secondly, numerous recent reports indicate that stromal fibroblasts and interstitial telocytes can also regulate functions of both parenchymal and barrier-forming tissues. These effects implicate diverse molecular mechanisms, including direct intercellular communication via homo- and heterocellular junctions, exchange with cellular constituents such as mitochondria, or by releasing regulatory cell factors within extracellular vesicles (e.g., exosomes). Moreover, a damaged tissue can promote migration of mesenchymal stromal cells (MSC) from the peripheral blood to the sites of injury, and stimulate their homing and release of factors sustaining aseptic environments and regeneration. These properties make MSCs and MSC-derived exosomes attractive for the cell therapy of diseases caused by radiation exposure, the radiation injury combined with secondary aggravating impacts such as trauma, and sepsis as well as for the acute viral infections. Recent developments in this area are also subjects of this book's discussion.

Perspectives and expectations from tissue barriers research

As our awareness of cellular mechanisms for tissue barrier regulation and reconstitution expands, so do the potential molecular targets for therapy of a variety of injuries and diseases. Then, it should always be remembered that an organism is prone to fail when barrier function fails. And, while separation of the effects due to the intrinsic cellular processes leading to the death of barrier-forming cells from the consequences of systemic hyper-reactive responses is not a simple task, cell therapy that is focused on barrier-forming targets is likely to be broadly applicable in future approaches for injury treatment.

Nikolai V Gorbunov, PhD

CHAPTER ONE

Roles and distribution of telocytes in tissue organization in health and disease

Carmen Elena Condrat[1,a] Mădălina Gabriela Barbu[1,a] Dana Claudia Thompson[1,a] Cezara Alina Dănilă[1], Andreea Elena Boboc[1], Nicolae Suciu[2,3] Dragoș Crețoiu[1,4], Silviu Cristian Voinea[5]

[1]Alessandrescu-Rusescu National Institute for Mother and Child Health, Fetal Medicine Excellence Research Center, Bucharest, Romania
[2]Division of Obstetrics, Gynecology and Neonatology, Carol Davila University of Medicine and Pharmacy, Bucharest, Romania
[3]Department of Obstetrics and Gynecology, Polizu Clinical Hospital, Alessandrescu-Rusescu National Institute for Mother and Child Health, Bucharest, Romania
[4]Department of Cell and Molecular Biology and Histology, Carol Davila University of Medicine and Pharmacy, Bucharest, Romania
[5]Department of Surgical Oncology, Prof. Dr. Alexandru Trestioreanu Oncology Institute, Carol Davila University of Medicine and Pharmacy, Bucharest, Romania

1. Introduction

Host tissue barriers arrange numerous external and internal interfaces evolutionary adapted to sustain morphogenesis, cellular and molecular integrity, metabolic homeostasis and endurance of organs and systems, as well as to proactively and reactively countermeasure environmental impacts and pathogen invasions. Barriers represent ensembles of cells of different histogenetic origins (e.g., mesenchymal, epithelial, endothelial, perivascular) that configure stroma/connective tissue and numerous specialized organs/systems, such as airways, gastrointestinal (GI) mucosa, vasculature, and skin. Evidently, the structural-functional singularity (substance) of these ensembles is orchestrated by complex cellular mechanisms (e.g., cell—cell contact interactions) and extracellular (paracrine, endocrine) signaling systems; however, the detailed aspects of these interactive communications are far from being understood. Naturally, stroma, internal epithelium, vasculature, endothelium, and skin are not

[a] These authors contributed equally to this work.

Tissue Barriers in Disease, Injury and Regeneration
ISBN: 978-0-12-818561-2
https://doi.org/10.1016/B978-0-12-818561-2.00001-1

static structural physical barriers, but rather shaped by morphogenetic forces and driven by interaction with tissue/organ parenchyma, as well as reaction to external impacts, e.g., trauma, heat, dehydration, etc. Emerging hub-function of telocytes (TCs) in the histogenesis and remodeling of the barrier network is intriguing and overwhelming. Moreover, upon accumulation of knowledge on the structural-functional interaction of barrier components across organs and systems, we expect further elucidation of integrative roles of TCs in tissue regeneration and engineering.

2. Telocytes—definition and history

TCs, previously known as Cajal-like cells (CLCs), are a recently discovered unique type of interstitial cells characterized by a small cell body and thin, long prolongations called telopodes (Tps) [1−3]. The history of TCs began in 1889 when Santiago Ramon Y Cajal first described what he called "interstitial neurons" in the muscle tunica of the gut [3]. He identified them using methylene blue and silver chromate staining which lead him to believe that they were primitive neurons [4]. Following his discovery, further ultrastructural studies have been made regarding the interstitial cells of Cajal (ICC), as they are now called, which have been shown to be distinct from neurons, fibroblasts, and smooth muscle fibers and to have roles in gut motility and neurotransmission [5,6]. From then on, such cells have been observed in different tissues outside the gut and given the morphological and immunohistochemical similarity, they were referred to as interstitial Cajal-like cells (ICLCs) [7−12]. It has been proven not only that ICLC and ICC are different types of cells that coexist in the interstitium but that they are also different from fibroblasts, neurons, fibrocytes, etc. [13,14]. Although having a distinct immunohistochemical profile, these cells share some markers with several other cell types (e.g., endothelial cells—CD34, ICC—c-kit, fibroblasts—vimentin, etc.). Particularly, positivity for vimentin, the major cytoskeletal component of mesenchymal cells [15], has led to the labeling of these cells as mesenchymal in origin [3,16]. Furthermore, to underline their unique structure and functions that differentiate them from other types of cells, in 2009, they were given the name "telocytes" by Popescu et al. [3]. The term telocytes refers to these cell's long and thin extracellular prolongations now known as Tps [1,3,12].

The discovery of TCs opened a world of possibilities in the medical field. Since then, many studies regarding TCs have been conducted to discover their functions in the human body and to identify possible clinical applications that would hopefully aid in the treatment of various diseases. These relatively new discovered cells seem to be involved in a multitude of physiological functions in both humans and animals, among them being cell signaling, tissue repair, the immune response, and probably many others that haven't yet been found [12].

To get an overview on the most researched subjects, we briefly reviewed articles published in 2018 and 2019 in the PubMed database containing the word "telocyte(s)" in the title. We excluded studies referring to CLCs and to ICC as they would not fit our view of the cell's particularities. We found 68 articles that met the mentioned criteria and observed a slight increase in the interest on this subject (Fig. 1.1). The most numerous articles on TCs over the last two years were published in relation to the reproductive system, followed closely by the digestive and cardiovascular systems and many others, as it can be seen in Fig. 1.2.

3. Localization and identification methods

Since 2009 when they were proved to be a distinct type of cells, TCs have been observed in various cavitary and noncavitary organs, in both humans and animal subjects. The known locations of TCs and related pathologies, along with bibliographic sources are summarized in Table 1.1. The data contained in Table 1.1 was also obtained from the PubMed database.

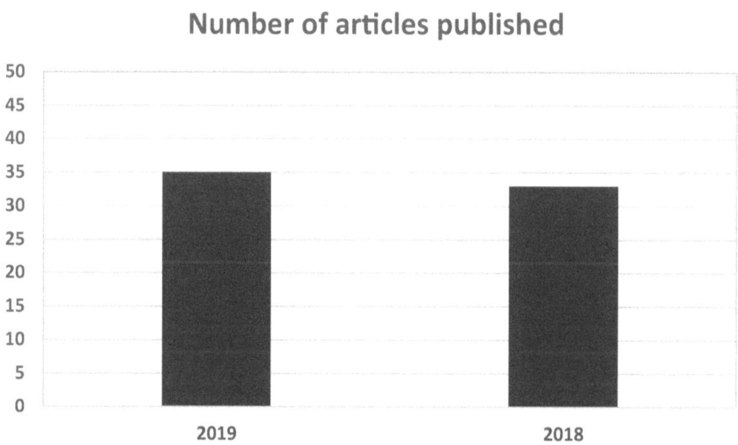

Figure 1.1 Number of articles on telocytes published in 2018 and 2019.

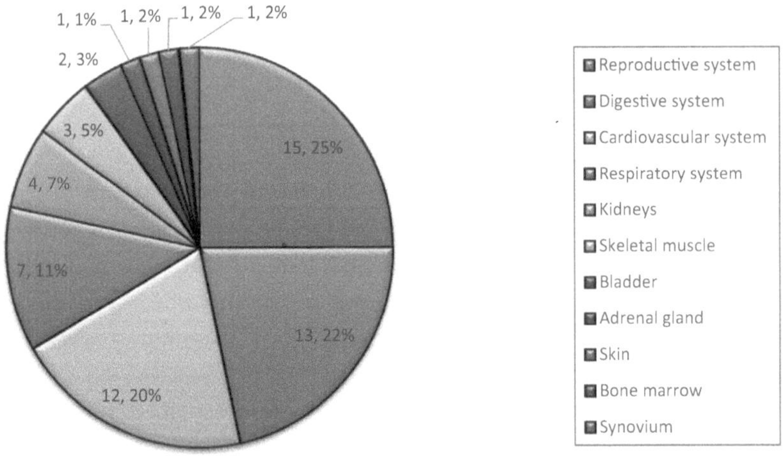

Figure 1.2 Articles on telocytes in various systems and organs.

Table 1.1 The localization of telocytes.

Location	Related pathologies/Primary role	References
Reproductive system		
Endometrium	Endometriosis	[17,18]
Myometrium	Leiomyomas	[19,20]
Ovaries	Chemotherapy-induced premature ovarian failure	[21]
Oviducts	Acute salpingitis; infertility	[22–24]
Mammary gland	Breast cancer	[25]
Placenta	Preeclampsia	[26,27]
Testis	Testicular seminoma	[28,29]
Prostate	Prostate cancer	[30]
Cardiovascular system		
Epicardium	Cardiac repair	[31]
Myocardium	Acute myocardial infarction	[32–34]
Endocardium	Blood-heart barrier	[35]
Sinoatrial node	Arrhythmias; atrial fibrillation	[36,37]
Heart valves	Mechanical support	[38]
Digestive system		
Tongue	Muscular trophism and activity regulation	[39]
Esophagus	Esophageal disorders involving angiogenesis and/or cell differentiation	[40,41]
Stomach	Regeneration and homeostasis regulation	[42]

Table 1.1 The localization of telocytes.—cont'd

Location	Related pathologies/Primary role	References
Jejunum	Homeostatic renewal of the intestinal epithelium	[43]
Colon	Ulcerative colitis; inflammatory fibroid polyps	[44,45]
Biliary system	Gallstone disease	[46,47]
Pancreas	Acinar cell regeneration	[48]
Liver	Liver fibrosis	[49–51]
Respiratory system		
Trachea	Tracheal secretion and contraction	[52]
Pleura	Barrier function	[53]
Lungs	Stromal support	[54,55]
Exocrine glands		
Small salivary glands	Ectopic lymphoid neogenesis; inflammation	[56]
Parotid glands	Homeostasis	[57]
Urinary system		
Kidneys	Renal fibrosis	[58,59]
Ureters	Tissue regeneration	[60]
Urinary bladder	Neurogenic detrusor overactivity	[61,62]
Skin	Systemic sclerosis; skin cancer	[63]
Skeletal muscle	Muscle repair	[64]
Bone marrow	Bone repair	[65]
Blood vessels	Homeostasis, angiogenesis and regeneration during vascular injury	[66]
Trigeminal ganglion	Microvascular maintenance and repair	[67]

3.1 Electron microscopy

There are two main types of electron microscopy (EM). The first one, transmission electron microscopy (TEM), is presently the gold standard for the identification of TCs and provides an enhanced two-dimensional image of their ultrastructure, while scanning electron microscopy (SEM) is able to produce a three-dimensional image [68]. While expensive, time consuming and demanding well-trained handlers, EM is currently the best method for TC visualization [68]. Recently, an improved technique called focused ion beam scanning electron microscopy (FIB-SEM) has emerged, which also provides three-dimensional (3D) images at nanoscale resolution [68] (Fig. 1.3).

TCs have a particular ultrastructure that distinguishes them from other cell types such as ICC, fibroblasts, etc (Table 1.2). They have a cellular

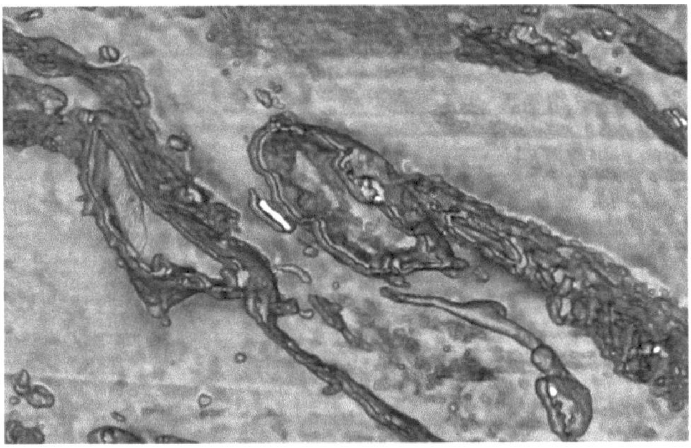

Figure 1.3 Three-dimensional reconstruction of telocytes in myocardium using focused ion beam scanning electron microscopy serial images [69]. *Source: Courtesy of Dr. Cretoiu D, Division of Cell Biology and Histology. University of Medicine, Bucharest, Romania and Carl Zeiss Microscopy GmbH, Munich, Germany; Cretoiu SM, Popescu LM. Telocytes revisited. Biomol concept 2014; 5(5):353–69.*

Table 1.2 The general ultrastructure of telocytes.

Cellular body	
Size	9–15 μm
Shape	Piriform, spindle, triangular, stellate
Nucleus	One, lined with heterochromatin; rarely visible nucleoli
Organelles	Mitochondria, Golgi apparatus, endoplasmic reticulum
Primary cilium	Inconstantly present
Extracellular vesicles	Yes—exosomes, ectosomes, etc.
Telopodes	
Number	1–5 per cell, usually 1–3
Size	Length = 10–1000 μm; thickness = 0.1 ± 0.5 μm
Appearance	Alternating thin and dilated segments—appearance of "bead-on-a-string"
Organelles	Yes—mitochondria and endoplasmatic reticulum—in the dilated segments (podoms)

body measuring only 9–15 μm [70], which varies in form depending on the number of Tps present, such as: when the TC only has one Tp, the cellular body is often described as piriform; when two, three or more Tps are present, the cellular body has a spindle, triangular or stellate form respectively [71]. Furthermore, the body contains just small amounts of cytoplasm around the nucleus, with sporadic organelles comprising mainly of mitochondria, endoplasmic reticulum, and Golgi apparatus [70,71]. The nucleus is frequently oval shaped, with a thin band of heterochromatin and rarely visible nucleoli [68]. According to some studies, TCs also present a primary cilium with uncertain functions, similar to ICC and other cells [52,68,72].

The cellular prolongations of TCs, known as Tps, are long (10−1000 μm) and thin (0.1 ± 0.5 μm) and are comprised of dilated segments called podoms, which alternate with thinner segments named podomers, conferring the overall appearance of "bead-on-a-string" [68,70,71]. Their number varies from one to five Tps per cell, with an average of one to three Tps/cell [70]. Organelles represented by mitochondria and endoplasmic reticulum are present in the dilated segments of the Tps as well as in the cellular body [13]. Through their Tps, TCs establish connections with different types of cells, such as fibroblasts and endothelial cells [71], but this is not the only type of intercellular communication. It has been shown that TCs release numerous extracellular vesicles (ECvs) like exosomes, ectosomes, and apoptotic bodies, thus passing valuable information to the adjacent cells [73].

Immunohistochemistry is a less reliable method for the identification of TCs, as it has been shown that the positive or negative reaction to different markers is variable through different tissues and organs [2]. This may suggest that TCs could have an organ-specific immunochemical profile. Furthermore, various studies have had different results regarding this subject even when they were focused on the same tissue/organ. For example, in the human myometrium, TCs were either c-kit positive [74], either c-kit negative but CD34 positive [75], while in other studies they were both CD34 and c-kit positive [70]. Other available markers for the identification of TCs are CD117, VEGF, vimentin, and caveolin-1, however, as stated before, the positivity to each of these markers is influenced by the source of the tissue sample processed [2].

4. Physiological functions of telocytes

Many hypotheses regarding the roles of TCs have been proposed since their discovery, yet none of them have been entirely proven to be true. Some of the existing theories suggest that TCs might participate in cell signaling [76], angiogenesis [77], organ regeneration and repair [2,64,78], apoptosis [79] and lately the possibility of them constituting a new form of tissue barrier has emerged and awaits to be validated. There is proof of the existence of TCs in different stages of organ development, so it is only logical that they would be involved in a multitude of crucial functions in the human body [80,81]. Furthermore, we will address each of the main theories concerning the roles that TCs play in the functioning of the human body, trying to explain the mechanisms through which they would exert them and present the latest data from literature supporting these premises.

4.1 Cell signaling

TCs have been identified in the stromal compartment of various tissues and organs, such as reproductive organs, mammary glands, urinary and GI tracts, heart and placenta, where they arguably possess many functions, with one of the most important being their participation in cell signaling [69,82]. Intercellular signaling is exerted through direct cell—cell interaction or through juxtacrine or paracrine interaction aided by shedding bodies and exosomes delivered by Tps. The election method for the precise detection of biological structures is FIB-SEM, which ensures an accurate visualization of cellular architecture by providing sharper images of 3D bodies, reaching nanoscale resolution [69,82].

TCs are implicated in long-distance communication through their Tps [82]. They also form a 3D network by establishing direct homocellular contacts between Tps and heterocellular junctions with other adjacent cells such as nerve endings, different kinds of progenitor cells, blood vessels and stromal cells including fibroblasts, mast cells, and macrophages [55,83—85]. Homocellular junctions ensure mechanical support and permit intercellular transfers that can be made through the simple apposition of plasma membranes or through complex contact sites. The mechanical role is accomplished through different kinds of adherens junctions that either connect overlapping Tps and are referred to as "puncta adhaerentia minima" and "processus adhaerens" junctions, or display a cuff-like appearance and are called "recessus adhaerens" or "manubria adhaerenta" junctions. Heterocellular junctions are made up of either minute contacts with a distance of only 10—30 nm between the two membranes or a simple apposition of the plasma membranes of TCs [83].

Furthermore, TCs have a significant impact on the activity of adjacent cells by the release of paracrine secretion mediated by ECvs, thus playing an important role in intercellular communication [70,86]. ECvs are made up of a lipid bilayer membrane that contains and transports different types of molecules in the extracellular space. TCs release diverse subtypes of

ECvs and the ones that have been studied the most are the exosomes. These are nano-sized vesicles that arise following the fusion between the plasma membrane and multivesicular bodies [87–91]. Exosomes have a diameter of about 30–150 nm and carry bioactive material such as miRNA, no-coding RNAs, cytokines, growth factors, chemokines [87–90,92]. Another specialized subtype of ECv are the ectosomes, with a diameter that varies between 50 and 1000 nm, formed directly from the cell membrane [73,87,90,91,93]. Ectosomes can carry glycoproteins, matrix metalloproteinases, and integrins. Some other large ECv are the multivesicular cargos (holding tightly packaged endomembrane-bound vesicles), which are lately considered to be released by cardiac TCs and the apoptotic bodies that are secreted upon apoptosis [73,92,94,95]. Through intercellular contacts and paracrine secretion, TCs ensure intercellular signaling, contributing to organ physiology, injury response, and tissue regeneration.

4.2 Tissue homeostasis, regeneration, and repair

Tisular homeostasis is a complex cellular process that involves the long-term maintenance of an internal balance within a given tissue through the regulation of cell proliferation and death and the control of metabolic networks and reactions. Stem cells need to divide and differentiate as a response to tissue injury while avoiding an excessive proliferation. Stem cell compartments are rather limited spaces that incorporate not only stem cells but also tissue infrastructure elements such as supporting and interstitial cells, extracellular matrix proteins like collagen, laminin and fibronectin, vascular and neural components. An adequate microenvironment ensuring that stem cells get to thrive is therefore crucial for them to reach their full regenerative potential. Because of the ability of TCs to communicate with one another and with other cells, both through the 3D networks that they establish and through paracrine signals, they may actually act as cell differentiation regulators and inductors. Moreover, once maturity is reached, TCs may have the capacity to differentiate as specific mature cells by acting as mesenchymal stromal cells that possess stem abilities. With the aid of currently available technology, it has been indicated that TCs cover a wide range of functions, stretching from their role in electrical conduction and contractility adjustment to the coordination of tissue regeneration and angiogenesis [69,81,96–98].

TCs were found in close proximity to or even within the stem cell niches of various organs such as the heart, dermis, meninges, bone marrow, and eye, with the supposition that they are implicated in tissue regenerative mechanisms [69,97,99]. For instance, in the heart, TCs are connected with cardiomyocytes through minute junctions featuring nanocontacts that facilitate the transmission of electrical signals [32,84,97,100,101]. TCs are placed in cardiac valves, epicardium, endocardium, myocardial interstitium, and they were also found in epicardial stem cell niches. They regulate stem cells directly by intercellular junctions and indirectly by releasing ECv. The various subtypes of ECv released by TCs are also implicated in the transfer of genetic material, such as microRNAs. It was demonstrated that cardiac TCs carry ECvs loaded with miRNAs to cardiac stem cells, the exchange being bidirectional, with cardiac stem cells transporting vesicles loaded with miRNAs to TCs, thus making an important contribution to cardiac homeostasis maintenance [94,97]. Moreover, TCs have been shown to express proangiopoietic miRNAs such as miR-100, miR-130, and miR-126, and their delivery through extracellular microvesicles to intended cells can help regenerate the damaged tissue by promoting angiogenesis. This supposition is further sustained by the fact that the expression of these miRNAs after acute myocardial infarction (MI) is raised [92].

Many studies demonstrated that TCs have different gene expression and mRNA signatures that make them highly distinctive from fibroblasts [69,102,103]. Specific genes are known to be needed for both tissue homeostasis and optimal stem cell function. To this extent, the overexpression of Calpain2 (Capn2), Four and a half LIM domains 2 (Fhl2), and quiescin Q6 sulfhydryl oxidase1 (QSox1) genes in TCs leads to speculation regarding their involvement in both morphogenesis and tissue homeostasis maintenance. These genes were found to be 100 times upregulated on chromosome 1 of pulmonary TCs when compared to fibroblasts [103]. The Capn2 encodes the calpain-2 catalytic subunit, a protein that participates in embryogenesis and cell migration in response to calcium signaling. Fhl2 is a regulator in various signaling pathways, playing the role of transcriptional coactivator after nuclear translocation by interacting with plasma membrane. QSox1 is implicated in extracellular matrix remodeling, cell cycle management, and oxidative protein folding [102].

The idea of using stem cells and progenitor cells as a therapeutic strategy because of their regenerative roles has been studied for many years. Stem cell niches contain constructive elements such as interstitial supporting cells, blood vessels, extracellular matrix proteins, and neural inputs [104]. Solid evidence has suggested that TCs exist in organs and tissues grouped in stem cell niches which are part of an interstitial network of high complexity that includes nerve endings, blood vessels, resident stem cells, and other interstitial components, thus bringing a critical contribution in regeneration and tissue repair [55,64,105–107].

Fibroblasts along with TCs are considered resident cells of the connective tissue. While fibroblasts play an important role in mechanotransduction and in various other processes such as inflammation and tissue remodeling, TCs have been discovered to be significantly involved in mechanosensitivity. At the same time, they have regenerative and reparative properties because of their long Tps that connect numerous cell types, thus restoring their functionality [108]. Moreover, the elongation, ramification, and deviation of Tps may modulate the function and morphology through establishing heterocellular and homocellular interactions [109]. Homocellular junctions (Fig. 1.4) can either be adherens or gap junctions. Adherens junctions have mechanical functions such as stabilizing the network, while gap junctions are responsible for intercellular signaling between TCs. Heterocellular junctions can also be found between TCs and many different cell types [55,70,86,110,111]. It is suggested that these spatial relationships with various cells might be significant for the implication of the TCs network in coordination and integration of different information regarding tissue homeostasis as a response to local functional demands. Moreover, TCs have the ability to transmit bidirectional information by means of ECvs, thus playing a modulator role in stem cell differentiation and proliferation [94,112,113].

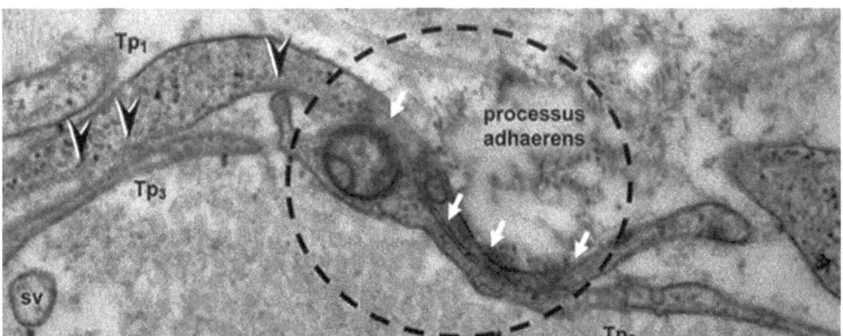

Figure 1.4 Electron microscopy image shows homocellular junctions in human heart: telopodes Tp 1 and Tp 2 are connected through puncta adherentia minima (*small arrows*) within a processus adhaerens. Small adjoining points of the telopodes' plasma membrane can also be observed (Tp 1–Tp 2 and Tp 1–Tp 3, *black arrowheads*) [110]. *Source: Gherghiceanu M, Popescu LM. Cardiac telocytes—their junctions and functional implications. Cell Tissue Res 2012; 348(2):265–79.*

There is a hypothesis regarding the ability of TCs to undergo phenotypic changes based on the functional needs of specific tissues. These phenotypic shifts explain the differently expressed markers that respond to various received signals. It has also been suggested that TCs, because of their ability to auto-regulate their functions and to transmit and receive signals from damaged tissue, participate in tisular regeneration and integrity maintenance [114]. Moreover, researchers have also confirmed that TCs are actively involved in stretch sensing, neurotransmission and GI motility. For instance, fibrosis correlated with a reduction in the number of TCs has been observed in the myenteric plexus and muscularis propria of patients with Crohn disease. At the same time, a decrease in TCs number and a fibrotic remodeling in the intestinal wall was observed in this condition [115].

In the heart, TCs were reported in epicardial groups of stem cells in contact with nerve terminations, resident cardiomyocytes precursors, and blood capillaries, providing support in the interstitial stroma for stem cells and cardiomyocytes progenitors to encourage continuous heart regeneration [116]. An increased number of TCs was reported in the peripheric area of heart ischemia during the neo-angiogenesis stage that follows the MI. These TCs were reported to establish connections with the endothelium of neo-formed and pre-existing capillaries [34]. The contribution to neo-angiogenesis is possible because of their paracrine secretion, which is immunohistochemically demonstrated by their positivity for vascular endothelial growth factor (VEGF) and nitric oxide synthase 2 (NOS2). TCs also play a role as "nursing" cardiomyocyte primogenitors in stem cell niches inside the epicardium and are therefore active participants in cardiac renewal. Moreover, TCs are supposed to induce and stimulate cell proliferation while inhibiting apoptosis and interstitial fibrosis. Situated in close proximity or even within stem cell niches, TCs have a tight relationship with stem cells, by stimulating, nurturing, and communicating with them [117].

4.3 Tissue barrier

Tissue barriers safeguard key physiological processes in tissues of the corresponding organs through restrictive paracellular and/or transcellular diffusion via tight junctions. Many of these tissue barriers are not static ultrastructures. Instead, these junctions undergo dynamic changes, requiring continuous disassembly, reassembly, and stabilization/maintenance [118].

All body surfaces and cavities are lined by layers of epithelial cells, which are connected by cell–cell junctions. These junctions serve three main purposes:

- adhesion to maintain tissue integrity,
- creation of a barrier to control the passage of ions, water, molecules, cells, and pathogens across epithelial layers,
- signaling to receive and transmit cues that affect cell behavior and tissue function [119].

The remodeling of tissue barriers is under tight physiological control, involving signaling pathways which facilitate the remodeling (i.e., opening, closing, and stabilization) of the barrier. This also prevents unwanted exposure of an organ or tissue to pathogens and/or harmful substances. The complex networks formed by TCs have been proposed to create potential barriers between tissues and parenchymal compartments. This theory is supported by the abundant collagen fibrils that surround the Tps, thus strengthening their pivotal role in the establishment of tissue architecture and control of paracellular and/or transcellular diffusion.

4.4 The placental barrier

The human placenta is an elaborate organ that ensures the normal evolution and maturation of the fetus by constantly changing throughout the pregnancy. The proper functioning of the placenta is achieved through the equilibrium between proliferation, differentiation and cellular death, and homeostasis maintenance is desirable especially keeping in mind that pregnancy is, in fact, a state where this balance can easily be harmed [120]. Four complex layers contribute to the formation of the placental tissue barrier:

- capillary endothelium of the villus,
- loose connective tissue that surrounds the endothelium,
- cytotrophoblast,
- syncytiotrophoblast.

At the end of the first week of gestation, blastocyst implantation occurs and the placenta begins to develop. At the time of implantation, the syncytiotrophoblast and cytotrophoblast originate from throphoblast, the outer region of the blastocyst, and are situated in the primary villi of the placenta. In the second week, gastrulation begins, during which the mesoderm is formed. The extraembryonic mesoderm is inserted into the primary villi, creating the secondary villi and, in the third week of gestation, the blood vessels start forming in all places where the mesoderm is developed, developing the tertiary villi, where TCs originate from mesenchymal cells [121].

TCs were identified among the multitude of cells in the placental stroma, which include fibroblasts, filamented cells, Hofbauer cells, vacuolated cells, and mesenchymal cells. TCs are known to regulate the placental function and the structural characteristics of the villi in the intervillous space by cell—cell interaction and paracrine signaling molecules. Placental TCs are placed between myofibroblasts and smooth muscle cells of the blood vessels, while possessing pacemaker properties somewhat similar to those in the intestine [122,123]. In studies performed by Suciu and colleagues, placental TCs were observed to make intimate contacts with Hofbauer cells, mast cells, myofibroblasts. They also confirmed that these TCs were positive for VEGF, caveolin-1, CD34, inducible nitric oxide synthase (iNOS), vimentin, and c-kit. In addition, in cell cultures, placental TCs were reported to make up to 14% of the cellular population, following fibroblast-like cells, which represent 51%, and smooth muscle cells, with a percentage of 25% [124,125].

Moreover, it was found that there are significant collagen deposits near the Tps of placental TCs [126]. When TCs of the chorionic villi die during apoptosis, Tps are able to modify their structure, therefore leading to damages in the synaptic-like connections between TCs and smooth muscle cells from fetal blood vessels, between TCs and myofibroblasts and between TCs themselves [127]. TCs therefore lose their pacemaker abilities because of the degradation of chorionic villi contraction/relaxation cycle. This may lead to a decrease in maternal metabolic input to the fetus by decreasing the contact area in the intervillous space, between the villi and the maternal blood [127].

4.5 The intestinal mucosal barrier

The GI tract is made up of structures that regularly change their diameters in concordance with food intake. Its complexity lies in its layers, which include the epithelium, lamina propria, and muscular mucosae, while the submucosa lies underneath. The motile activities, peristalsis, and relaxation/contraction cycle are ensured by the muscular layer, while the enteric nervous system and pacemaker cells are responsible for coordination. Intestinal TCs create a 3D network in different layers of the gut, surrounding the blood vessels, intestinal crypts, gastric glands and nerve fibers, muscle layers, submucosa, and mucosa. The complex web that they form has an important structural function since it virtually creates a scaffold which directs the collagen bundles and fibers, allowing for the arrangement of spaces that hold the connective tissue components [128], as illustrated in Fig. 1.5 [129].

The intestinal mucosal barrier is a heterogenous entity composed of physical, biochemical, and immune elements elaborated by the intestinal mucosa. It is made up of the mucus layer, the intestinal epithelium, and the microbiota [130]. The intestinal epithelium is continually renewed by stem and progenitor cells, and its turnover rate is the highest of all human body tissues, being replaced in 3–5 days [131]. The proliferation takes place

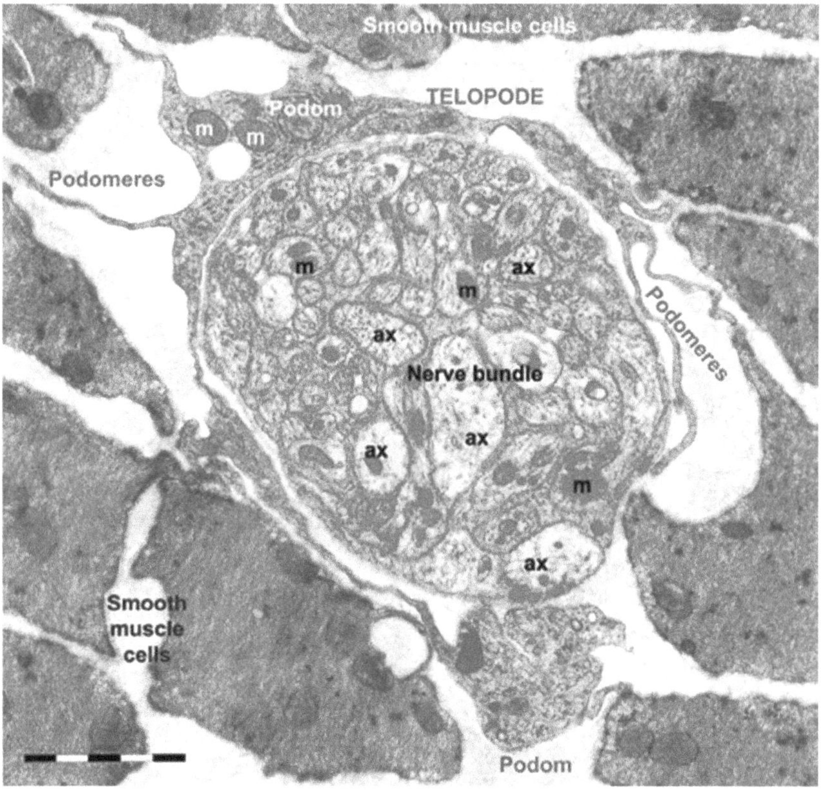

Figure 1.5 Muscularis mucosa of rat jejunum. A color-enhanced digital micrograph of a black and white transmission electron microscopy image. A telopode (blue [dark gray in print version]) is pictured surrounding a nerve bundle (green [gray in print version]) between smooth muscle cells (brown [black in print version]); *ax*, axon; *m*, mitochondria [129]. *Source: Cretoiu D, Cretoiu SM, Simionescu AA, Popescu LM. Telocytes, a distinct type of cell among the stromal cells present in the lamina propria of jejunum. Histol Histopathol 2012; 27(8):1067–78.*

in selected areas, called intestinal crypts, which extend to the underlaying stroma, and it is activated because of signals and growth factors released by local niche cells.

In the intestinal muscular layer, TCs' prolongations create 3D webs interfusing with processes of the ICCs and make cell-to-cell contacts visible by immunohistochemistry and EM. The networks formed by TCs can be distinguished by those formed by ICCs because of their distinct immunophenotype. TCs are negative for c-kit and CD34+/PDGFR-α+, as opposed to ICCs, which are positive for c-kit and negative for PDGFR-α or CD34 [132], as illustrated in Fig. 1.6 [133].

These networks indicate that intramuscular TCs might support the dispersal of the slow waves produced by ICCs, which are making electrical

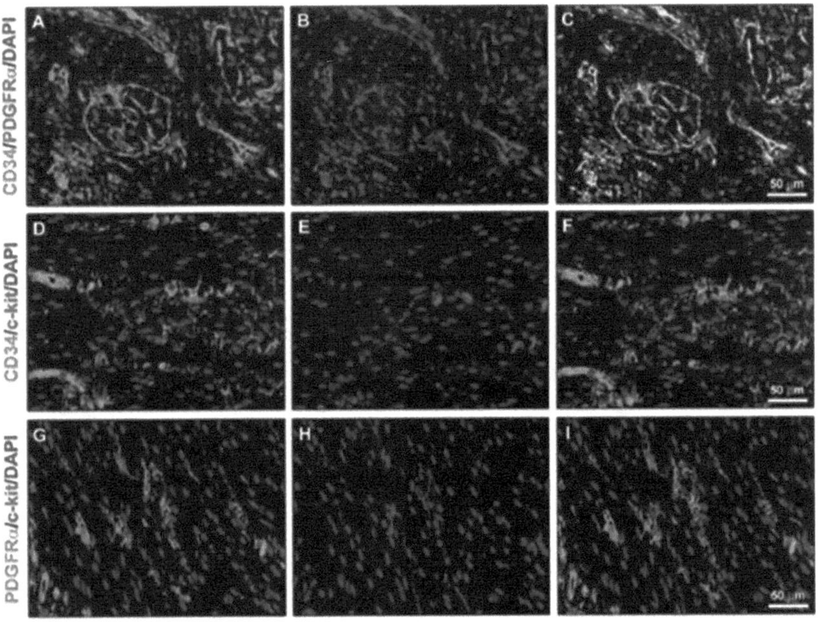

Figure 1.6 nterstitial cells of Cajal and telocytes in the muscularis propria of the human gut that form interlocking networks. Human colon sections double immunostained for: (A–C) CD34 (green [gray in print version]) and PDGFR-α (red [dark gray in print version]), (D–F) CD34 (green [gray in print version]) and c-kit (red [dark gray in print version]), and (G–I) PDGFR-α (green [gray in print version]) and c-kit (red [dark gray in print version]). Nuclei are counterstained with DAPI (blue [black in print version]). The right panels show the merged images. All telocytes are CD34+/PDGFR-α+ (A–C), while ICCs are c-kit+ and CD34 negative (D–F), PDGFR-α negative (G–I) [133]. *Source: Cretoiu D, Vannucchi M, Bei Y, Manetti M, Faussone-Pellegrini M, Ibba-Manneschi L et al. Telocytes: new connecting devices in the stromal space of organs. 2019.*

contact with smooth muscle cells, thus playing an important role in the motility of the GI tract [132,134]. Moreover, taking this observation in consideration, it was indicated that smooth muscle cells are electrically coupled not only to ICCs but also to PDGFR-α+ cells resembling TCs, thus setting up a unit referred to as the SIP syncytium, deeply involved in GI motility because of its pacemaker activity [135]. Based on the intertwined TCs/ICCs networks, one interesting hypothesis is that TCs localized in the muscular layer of the GI tract might convert into ICCs, however, further studies are needed to demonstrate this theory [78,136].

Subepithelial TCs have been recently found to be stamped by the expression of the winged helix transcription factor Fox1 and by hedgehog signaling mediator Gli1. These TCs form a continuous plexus underneath the intestinal epithelium and send pro-proliferative Wnt signals to local stem and progenitor cells [137]. The importance of Fox1-expressing TCs in crypt maintenance was revealed in experiments performed on mice that were published in 2016. In these studies, diphtheria toxin mediated ablation in Foxl1-DTA mice not only stopped the epithelial proliferation but also it was followed by a shortening of the intestine, half reduction of the jejunal villus length and cutting by 95% of the cycling cells number per crypt in just a few days after the loss of Fox1+ TCs [138].

Foxl1/Gli TCs play an important role in regeneration and differentiation of the intestinal epithelium. These TCs are strategically distributed, so that the expression of pro-proliferative factors including R-Spondin and Wnt2b is increased in the intestinal crypts, while the expression of pro-differentiation factors such as BMP5 and Wnt5a is increased in the villi [139]. Because of the exceptional position of Foxl1/Gli TCs, firmly apposi-tioned to the intestinal epithelium, it is a future desire to understand and demonstrate that these TCs may interfere in the interaction of the immune system with luminal antigens after an injury of the intestinal barrier.

4.6 The skin barrier

The epidermis provides protective functions against environmental factors, its barrier effect arising as a result of the combined participation of the epidermal and dermal layers, with a main contribution coming from the stratum corneum [140]. These functions include permeability regulation, antimicrobial activity, photoprotection, as well as antioxidant effects [141].

Depending on location, the epidermis consists of four or five cell layers. The most profound layer, stratum basale, is made up of a single row of keratinocytes lying on the basement membrane, continuously multiplying while moving to the top to gradually restore the layers above, during the keratinization process [142]. It is situated just above the dermis, which offers a supportive matrix for the skin, consisting of a lean papillary layer joined to the epidermis and a dense reticular layer rich in collagen fibers [143]. Recently, TCs have also been identified preponderantly in the reticular dermis, around the perifollicular sheath, surrounding the glassy membrane that covers the hair root, around blood vessels, neighboring sweat and sebaceous glands and arrector pili muscles, possessing characteristics similar to those in other organs [106,144]. FIB-SEM analysis revealed the 3D-configuration of TCs in the human dermis (Fig. 1.7), with computer-assisted reconstruction of sections showing volumetric features of TCs and their Tps [145]. However, immunohistochemistry currently remains one of the most fitting methods to identify skin TCs, with Manetti and colleagues managing to differentiate this population from endothelial cells by defining them as cells positive for CD34 and negative for c-kit, CD11c, CD31, CD90, and α-SMA [146].

The unique configuration of skin TCs renders them as key players in the local architectural network, with their impairment leading to loss of organization of collagen and elastic fibers. Their participation in the local microenvironment is further supported by their paracrine secretion and transfer of signaling molecules enclosed in shedding vesicles, through which they adjust the functions of nearby fibroblasts, immunoreactive cells

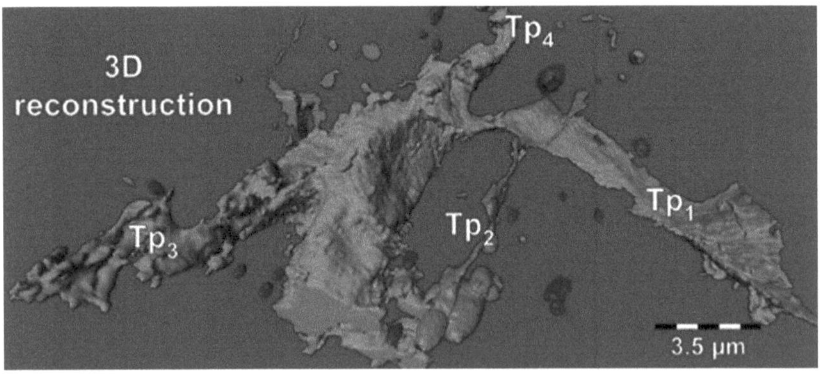

Figure 1.7 Focused ion beam scanning electron microscopy tomography revealing three-dimensional details of telopodes (Tp_1–Tp_4) and extracellular vesicles (purple [black in print version]) [145]. *Source: Cretoiu D, Gherghiceanu M, Hummel E, Zimmermann H, Simionescu O, Popescu LM. FIB-SEM tomography of human skin telocytes and their extracellular vesicles. J Cell Mol Med 2015; 19(4):714–22.*

and stem cells [146]. Moreover, their proximity to stem cells predicts their profound involvement in epidermal turnover and regeneration by their acting as nurse cells [106], fact further supported by the association between damaged TCs and defective epidermal cell turnover observed in psoriasis [147].

Evidently, the multiple roles of skin TCs certainly require more attention to further elucidate their involvement in skin homeostasis, remodeling and repair, as well as their contribution to various dermatoses and skin tumors.

5. Roles of telocytes in disease

TCs have been found recently in many human organs and have been shown to participate in various processes such as cellular aging prevention, inhibition of oxidative stress, and angiogenesis stimulation [34,69,148]. TCs are also participating in the process of immune surveillance and immunomodulation, standing as "local data suppliers" for the immune response, while also displaying electrophysiologic properties, as recent studies have demonstrated [9,56,69,129].

Furthermore, TCs are known to participate in the mechanism of intestinal mobility and neurotransmission in the enteric neuromuscular compartment, likely by spreading the slow-paced waves generated by the pacemaker ICC [1,13,69,132]. In heart disease, the dynamics of TCs dependably follows any changes in the quantitative and qualitative composition of the extracellular matrix [149], and studies have shown that cardiac TCs are decreased during heart failure [150]. It has been implied that a decrease in TCs number can be linked to impaired intercellular signaling and atypical 3D arrangement in the myocardium [150].

Chronic inflammation develops as a result of the complex interaction between tissue-resident stromal cells, immune and nonimmune cells [151], eventually leading to the development of fibrosis. It has been demonstrated that patients suffering from chronic inflammatory diseases such as Crohn's disease or ulcerative colitis have a decreased number of TCs causing dysmotility and architectural disorder [44,115]. This reduced number of intestinal TCs may have different reasons and different pathological implications. The consequent inclusion of TCs in the fibrotic scar combined with the gradual alteration of the extracellular matrix might cause deep cell damage, as suggested by human heart conditions [150].

TCs were also identified in the human myometrium and endometrium interstitium, as well as in the fallopian tube. The location of TCs in the myometrium may suggest their implication in myogenic contractile mechanism during embryo implantation, delivery, and sperm transport [17,86,152].

Owing to extensive distribution of TCs all over human body systems and many relations assigned to these cells, a significant number of studies addressed their implication in different disorders so as to discover different perspectives regarding potential treatment strategies [56,69,150].

5.1 Telocytes in breast cancer

Worldwide, one of the main causes of cancer mortality in women is breast cancer. During the female lifetime, there are many factors that influence breast tissue, including breast growth, menstrual cycle, pregnancy, lactation, and postmenstrual regressions. An early first menstrual period combined with nulliparity pose an especially high risk of developing mammary cancer [153]. However, on the other hand, genetics also play an important role, with hereditary ovarian and breast cancer being caused by mutations of the BRCA1 and BRCA2 genes. Identifying the presence of these mutations leads to significant decisions in terms of pharmacological options and surgical risk. Still, these genes can also be involved in the development of nonhereditary, sporadic tumors which incorporate somatic versions of the BRCA1 and BRCA2 mutation genes [153,154].

Epithelial cells and the associated stroma make up the mammary gland. The epithelial cells form the basal myoepithelial and luminal epithelial layers, while the mammary gland stroma includes adipocytes, macrophages, fibroblasts, TCs, nervous endings, and endothelial cells. These cells secrete substances with signaling roles in the stroma which participate in the development and normal function of the mammary gland. On the other hand, tumor stromal cells and cancer cells secrete numerous types of chemokines which encourage cell migration, dissemination, and tumor growth [155]. Mammary TCs are positive for CD10, vimentin, and CD34, and, as connective cells that offer mechanic support, significantly contribute to creating a supportive framework within the mammary gland.

Any alteration in the arrangement of the stroma and in the function of TCs inadvertently increases the risk of developing a neoplastic process in the mammary gland. This was demonstrated in 2012, when Mou et al. found that TCs form membrane-to-membrane contacts with adjacent breast cancer cells and presumably contribute to the formation of neoplastic cell clusters. Moreover, their study, which used two lines of breast cancer cells, namely stromal cells from normal mammary gland and reconstituted breast cancer tissue, showed a reduction in the apoptosis ratio and an increase in the proliferation index in breast cancer cells accompanied by stromal cells including TCs. In addition, they evaluated the roles of TCs within the extracellular matrix and intercellular junctions by evaluating the expression of E-cadherin, F-actin, and collagen IV, thus discovering that TCs encouraged the self-assembly of common breast structures. The spreading of TCs along cellular membranes as well as membrane to membrane contacts between them and with adjacent cells was observed via TEM. In breast cancer tissue, cell to cell communication through transmission of multivesicular bodies and gap-junctions has important contributions to the adaptation of cancer cells to local low pH, acidosis and hypoxia, conditions present in the tumor microenvironment [156]. It is hence of paramount importance to gain as much insight as possible into the stroma-tumor cell interactions to better understand the mechanisms involved which could help uncover possible therapeutic strategies against breast cancer.

5.2 Telocytes in uterus and fallopian tubes

The female reproductive system is made up of the external and internal reproductive organs, with the latter including the vagina, uterus, fallopian tubes, and ovaries. In the last decades, impressive progress has been made in analyzing the microanatomy and physiology of the uterus and fallopian tubes, keeping in mind their fundamental importance in reproduction and the bearing of offspring. Still, uterine contractility continues to pose significant challenge in numerous disorders, including infertility, spontaneous abortion, and premature birth [157]. The uterus consists of two smooth muscle layers: the myometrium (external) and the endometrium (internal), which, in turn, is made up of three morphologically distinct layers: the outer layer (stratum compactum), the intermediate layer (stratum spongiosum), and the inner layer (stratum basalis). The fallopian tube or the oviduct is composed by the external serosa consisting of loose supporting tissue, the intermediate smooth muscle layer, and the inner mucosal lining. In 2005, a distinctive cell type was described within these structures, which were speculated to play a pacemaker role [10,158] and were later identified as TCs [3].

Myometrial TCs tested by immunofluorescence are positive for CD117 (c-kit) in various areas of the membrane, while positivity for CD34 can also be identified. On the other hand, Tps are highly positive for vimentin, a cytoskeleton protein, and negative for α-smooth muscle actin (α-SMA) [158]. Further studies have concluded that the expression of PDGFRα/CD34 is also a characteristic of TCs in general [132,159]. Immunocytochemistry reveals intense positive nuclear staining for estrogen receptors (ERs) and progesterone receptors (PRs). Double staining for CD117 and ER as well as CD117 and PR further confirms these findings [160]. Therefore, it is currently established that myometrial TCs are double positive for CD34 and PDGFR-α while also expressing positivity for ER and PR. Endometrial TCs show positivity for CD34 and vimentin, as well as for connexin 43 [17,161]. Connexin 43 belongs to the connexin protein family, which are fundamental for the formation of gap junctions [162]. It has been shown that Connexin 43 in the stromal cells of the endometrium is essential for pregnancy-associated angiogenesis, decidualization, and subsequent embryo survival [163], while its reduced expression is linked to recurrent miscarriage [164], thus making it reasonable to speculate on TCs' involvement in successful term pregnancies.

In the fallopian tubes, TCs amount to about 8% of the thickness of the lamina propria and 7%–9% of the muscularis [158]. C-kit and vimentin positive tubal TCs are frequent in the mucosa, along with cells positive for CD34. While c-kit is mainly present on the cell body, CD34 is rather expressed on the Tps. Furthermore, Tps also express caveolin-1 and caveolin-2, proteins that aid the formation of caveolae, membrane structures involved in endocytosis, cell signaling, and cholesterol metabolism [165]. Tubal TCs are also known to express ER and PR. Both the cytoplasm and the prolongations of TCs express CD117, S-100 protein, desmin, which regulates muscle integrity and architecture [166], SMA, which plays an important role in fibrogenesis [167], laminin, a fundamental protein of the extracellular network [168], as well as tubulin [169], which governs the complex dynamics of cellular microtubules [170].

Intra- and subepithelial TCs are connected with each other through single or multiple Tps as well as fenestrated membranes, with the latter forming a continuous network along the muscular wall [169]. TPs intersect in multiple points, not only within the same layer but also with Tps of TCs from adjacent layers, thus developing a complex biological network that facilitates cellular communication. Furthermore, TCs are also linked to other types of cells, such as epithelial, mast cells, or vascular smooth muscle cells. On this note, their role in the transmission of slow waves within the muscular wall of the fallopian tubes has been observed [171].

However, while in the GI tract [172], urinary tract [173], or vascular tissue [72], TCs are believed to play a pacemaker role, in the uterine walls, because of the alternation between periods of contractility and relaxation, the need for such a pacemaker can be viewed as lacking [174,175], which might explain the decreased number of TCs in the pregnant uterus [176]. However, the modulation of contractility is also hormone dependent, which establishes the relevance of the expression of steroid hormone receptors on uterine TCs [177].

During pregnancy, the myometrium associates a relatively small number of TCs, as opposed to the postpartum uterus. Similarly decreased numbers of TCs are seen in the immature uterus. This attests to their role in controlling the bleeding and postpartum uterine involution because of their electrical properties that render them capable of regulating smooth muscle contractions [157,176]. Meanwhile, both the immature and the pregnant uterus have weak contractile activity, for the latter to prevent preterm labor [178,179]. Furthermore, uterine TCs can activate peritoneal macrophages, which in turn determine an increase of interleukins IL-1, IL-6, IL-10, tumoral necrosis factor-alfa, and iNOS, that can cause endometriosis, implantation failure, and immunologically mediates abortion [180].

The fallopian tubes of patients suffering from endometriosis or tubal ectopic pregnancy have been shown to be associated with the deterioration of TCs, which end up establishing a rather poor network with fewer connections. This damaged network further leads to impaired tubal motility [181]. In acute salpingitis models, TCs have been shown to suffer ultrastructural damage affecting the cellular body and the Tps, which ultimately leads to the deterioration of the 3D network. TCs injury is therefore considered to be partly responsible for the resulting tissue fibrosis because of their inability to successfully perform their regenerative function [23].

5.3 Heart disease

In the human adult heart, TCs have been described in the epicardium, myocardium, endocardium as well as within the local stem cell niches [35,96,101,116,182]. They have been shown to express CD117, CD34, PDGFR-β, and vimentin [31,183,184]. Following a study performed by Bei et al., double immunolabeling has been recommended to accurately differentiate cardiac TCs from fibroblasts, which are only positive for PDGFR-β and vimentin [185], and pericytes, which are CD34-negative/PDGFR-β positive [186]. The distribution of TCs within the cardiac layers is not homogenous: in the endocardium and myocardium, TCs are organized in cellular networks [187,188], while subepicardially they reside in the local stem cell niches [116,189]. The 3D networks that they establish aid the communication with surrounding cells, including cardiomyocytes, endothelial cells, fibroblasts, pericytes, etc. [190]. It is currently considered that cardiac TCs act together with local stem cells to enhance cardiac regeneration and repair. Increasing evidence points to the nursing and guiding roles of TCs toward stem cells [116,184]. To this extent, TCs promote cardiac repair by means of their close contact with both stem cells and cardiomyoblasts and through the release of exosomes [191,192]. Moreover, cardiac TCs can also promote the proliferation and differentiation of cardiac stem cells through paracrine mechanisms involving their secretome, which consists of higher levels of IL-6, VEGF, macrophage inflammatory protein-1α (MIP-1α), MIP-2, and MCP-1 [113].

A recent study has shown that cardiomyopathy caused by dilation, ischemia, or inflammation is associated with a reduced number of TCs. TCs and Tps were decreased or even absent where severe fibrosis was observed [150]. Furthermore, heart failure caused by MI also associates TC apoptosis [193]. A gradual reduction in the number of TCs can be witnessed in the myocardium affected by systemic sclerosis because of the widespread fibrosis [194], but also in the aging heart, accompanied by progressive tisular degeneration [101]. The exact causes of TC apoptosis in the diseased heart have not been clearly established. It has been demonstrated, however, that oxidative stress plays a negative role in TCs formation [195], while a similar negative impact can result from an altered composition of the extracellular matrix, as seen in scarring fibrosis after MI [193]. On the other hand, TCs can persist despite the aggression, but with ultrastructural deterioration consisting of cytoplasmic vacuolation and Tps shortening that renders them dysfunctional [190].

5.4 Pulmonary disease

The lung tissue consists of a heterogeneous population of cells that work together with various cytokines, cellular receptors, and signaling molecules to perform complex pulmonary functions such as gaseous exchange and immune response. TCs have been identified in the interstitium of distal bronchioles, where their Tps are in contact with alveolar epithelial cells [54]. Moreover, they can also be observed among smooth muscle cells and capillary endothelial cells [55]. While fetal lung TCs are more likely to associate larger cell bodies that surround the neocapillaries, adult lung TCs boast longer Tps full of secretory vesicles [80]. Lung TCs have been shown to express CD34, c-kit and vimentin, connexin 43, as well as VEGF, EGF, and neuron-specific enolase [80,196].

Lung TCs, because of their ability to form a resistant yet flexible 3D network, are believed to ensure mechanical support by maintaining open the lumen of bronchioles and blood capillaries during breathing movements [55,197]. Moreover, TCs may contribute to the air-blood barrier because of their close contact with interstitial capillaries [77]. The role that lung TCs play in immune surveillance is supported by their association with lymphocytes, macrophages, eosinophils, and basophils [9,198,199], eliciting the assumption that they are important players in pulmonary infectious diseases. Similar to heart TCs, lung TCs are situated in close proximity to stem cells, even establishing intercellular junctions, which led to speculation regarding their roles as guiding and nutrient cells for stem cells during tisular regeneration and reparation [55,80]. To this end, Zheng et al. have demonstrated that human lung TCs induced the production of VEGF and EGF, thus bolstering the proliferation of pulmonary microvascular endothelial cells in the process of angiogenesis [196].

Chronic obstructive pulmonary disease (COPD) is characterized by progressive airflow obstruction because of chronic inflammation that leads to a decline of lung function [200]. Because of their strategic localization between smooth muscle cells, their interaction with immune cells as well as their roles in tisular regeneration and repair, TCs are believed to be impaired in COPD, resulting in hyper-remodeling and the maintenance of a persistent inflammatory state [201]. Interstitial lung disease often occurs as a result of an abnormal healing response to lung injury [202]. It is currently thought that aberrant activity of TCs contributes to the process of interstitial fibrosis because of their ability to promote fibroblast proliferation [197,201]. During pulmonary infection, microbial stimuli can activate TCs, which in turn transmit intercellular signals to recruit immune cells from the bloodstream [201].

5.5 Digestive system

TCs have first been discovered in the digestive system and, throughout the years, have been shown to possess a number of functions and properties that make them act as tissue barriers. According to the studies conducted on both mice and humans, they form extensive networks in the muscle, submucosa, and mucosa layers, and also around blood vessels, nerves, intestinal crypts, and gastric glands [128,132]. From an immunohistochemical perspective, all TCs, regardless of their location in the GI wall, are CD34+/PDGFRα+ interstitial cells [132]. However, they tend to accomplish different roles according to their place of origin.

The role of the TCs encountered in the submucosa and lamina propria is primarily structural. They form 3D networks by connecting among themselves, thus leading to the formation of a scaffold that can influence the location of the collagen fibers or bundles and therefore defining the spaces that will accommodate various components of the connective tissue. In the intestinal crypts and gastric glands, on the other hand, the proposed role of these TCs has been linked to the replication and differentiation of epithelial stem cells because of the fact that they secrete and produce numerous molecules [128,132,203], they closely interact with the "stem cell niches" [136,203], and they express the PDGFRα receptor, essential for organogenesis [128]. A recent study has demonstrated that the TCs network encountered at this level and the expression of PDGFRα play an important role in providing Wnt proteins, indispensable for the proliferation of stem cells in the intestinal crypts [43].

In the muscle layer, TCs could play two different roles. One of these is represented by their potential contribution to the motility of the GI tract [132,134]. According to the TCs' immunohistochemistry and the EM analysis, these cells have been shown to form 3D networks with the ICC [136]. At this level, the differentiation between the two cells is very clear, relying on their immunophenotypes (ICCs are positive for c-kit, but negative for PDGFRα and CD34, while TCs are exactly the opposite) [132]. This theory is also supported by a recent paper, which states that TCs, ICCs, and smooth muscle cells are all electrically connected, forming an entity called "the SIP syncytium" [135]. Another proposed role of the TCs in the muscle layer of the digestive system is the possibility to ultimately transform into ICC. The reasons behind this hypothesis rely on the earlier described networks between TCs and ICCs and the fact that a decrease in ICCs has never been observed, even though researchers have described ICCs undergoing apoptosis, but none going through mitosis [204]. Another argument for this theory states that, in vitro, CD34+ stromal cells gradually lose their immunophenotype and replace it with c-kit positivity, which is characteristic for ICCs [205].

TCs have been shown to play an important part in the development of a number of pathologies associated with the digestive system. For example, in the liver, they have been observed surrounding the sinusoids [159], and they are thought to be implicated in both the fibrosis and the regeneration of this organ [50]. Also, a lower number of TCs have been associated with an increased gallstone occurrence [206,207]. Further on, the involvement of TCs in some of the better studied pathologies of the digestive system will be discussed in more detail.

5.6 The role of telocytes in cholelithiasis

A number of factors have been held responsible for the formation of gallstones, such as cholesterol excess, pronucleating proteins, bile salts that are hydrophobic and an abundance of mucus that leads to the formation of gel in the gallbladder [208–211]. However, another very important factor contributing to and maybe even initiating the genesis of cholesterol stones is represented by the dysmotility of the gallbladder [209,212]. This could potentially be influenced by TCs. This type of cell has first been linked to the human gallbladder when GI stromal tumors originating from the gallbladder were discovered to contain ICLCs that were CD117 positive [213,214]. Subsequently, TCs were identified in the gallbladder of animals, such as guinea pigs. Their presence in guinea pigs was also associated with smooth muscle cells electrical binding [215], modulating the tonus of the gallbladder [216], distribution of spontaneous excitability and rhythmicity of the gallbladder [217], and also their receptor for cholecystokinin-A was found to be indispensable for the contraction of the gallbladder muscles [218]. In the human gallbladder, TCs have been found in the lamina propria and among groups of smooth muscle cells, either alone, or in small bundles of two or three cells [219].

The proposed mechanism through which TCs are thought to be influencing the motility of the gallbladder and the formation of gallstones is related to the release of exosomes or vesicles [220]. In healthy subjects, TCs appear to be in greater numbers than in patients diagnosed with cholelithiasis. Under normal circumstances, they up-regulate the pacemaking mechanism of the gallbladder by forming complex 3D networks through both homo- and heterocellular junctions, therefore creating the mechanical and functional basis for long distance intercellular signaling [64,110,191].

5.7 The role of telocytes in liver fibrosis

Liver fibrosis is a pathology usually associated with chronic liver damage or fatty liver because of obesity [221–223], which can eventually lead to cirrhosis—a major cause of mortality around the world [224–226]. There is currently no standard treatment for liver fibrosis, and no drug has proven its efficiency in human subjects so far [226–228].

Studies have shown that TCs in liver fibrosis tend to suffer a considerable decrease in number, of up to 27%–60%. This change could be followed by an alteration of the extracellular matrix organization and also by a lack of control regarding the activity of myofibroblasts, therefore leading to the formation of fibrosis [51]. Further on, prior research has shown that TCs are located in the close vicinity of hepatic putative stem cells, thus they could play a significant role in the proliferation of hepatocytes and liver regeneration [229]. However, further studies need to be conducted to establish whether the transplantation of TCs could be a useful therapeutic tool for the treatment of liver fibrosis.

5.8 The role of telocytes in ulcerative colitis

The pathogenesis of ulcerative colitis involves multiple relapses of chronic intestinal inflammation that progressively lead to extensive fibrosis along the colonic wall. The final result is represented by an increase in the stiffness of the colon, which is in an inadequate state to perform peristalsis or to resorb nutrients [230,231].

TCs have also been found in the wall of the colon, and a number of roles have been attributed to them, such as intercellular signaling, regulation of the intestinal movements, and structural support [1,3,13,132,220]. A prior study that has analyzed samples from resected left colon specimens taken from patients suffering from different stages of active ulcerative colitis showed there was a significant difference in the presence and distribution of TCs in relationship to all of the above. For instance, while in both early and advanced cases of ulcerative colitis, TCs numbers were reduced in the muscularis mucosae and submucosa, in the early cases, they tend to be preserved in the more outer layers, such as the myenteric plexus and the muscle layer. On the other hand, in advanced ulcerative colitis, where fibrosis had already reached these outer layers, the distribution of TCs was scarce or even absent around the smooth muscle fibers and the nervous plexus [44]. Such changes are thought to be associated with the genesis of fibrosis that is responsible for remodeling the colonic tissue and could also lead to an abnormal bowel movement [115].

6. Perspectives and conclusion

Even though TCs have been discovered for many years, their roles and potential therapeutic uses are still being elucidated. Perhaps one of the most promising perspectives that have been discovered so far is related to their involvement in the neoangiogenesis following MI [232].

Even though this pathology affects a major percentage of the population worldwide and the scars resulted leave the patients with various disabilities, there is currently no treatment that could lead to the regeneration of the post MI fibrosis. However, TCs have been shown to play an important role in nursing the stem cells in the myocardium and could offer a novel therapeutic approach for this ailment [116]. So far, two methods have been proposed. The first one involves the intramyocardial transplantation of TCs in the area of the infarction and also the surrounding zones. As a result of this experiment, two weeks after the procedure, the area of infarction had decreased significantly, and the left ventricle function had been improved [193]. The mechanisms are not yet fully understood, but some studies reported that TCs could express certain microRNAs associated with angiogenesis, secrete VEGF, and also connect with newly formed cells derived from the endothelium at the edges of the MI [34,110,116]. The second method used iPSC or "derived human mesenchymal stem cells" for transplantation in the infarcted area [233]. This procedure increased the number of TCs in the affected area, in addition to improving ventricular remodeling.

In regards to other ailments, TCs could also prove their usefulness in developing therapies for fibrosis in the digestive system. Future research should focus their efforts on the possibility that TCs could be in charge of preventing the activation of fibroblasts. Also, they possess the ability to diminish the chaotic organization of the extracellular matrix encountered during fibrotic processes [44]. In the reproductive system, experiments conducted on rats showed that in the oviduct affected by endometriosis, TCs were damaged, potentially affecting the stem-cell tissue repairing processes. In such cases, future therapies could target TCs alone or in combination with stem cells to prevent irreversible tissue damage and even to promote regeneration [18]. In primary cultures of pregnant myometrium low-level laser stimulation has been found to enhance the growth of TCs, thus leading to the hypothesis that future therapies should also focus on promoting the local growth of TCs, rather than solely on exogenous transplantation [105].

Besides these, other areas of medicine could benefit from the regenerative potential offered by the TCs and their close relationship with stem cells. Such areas include pulmonology, where lung regeneration could be achieved through TC-progenitor cells communication via exosomes or close nano-contacts [55], dermatology for skin regeneration and repair [106], neurology [234], and potentially many more [2].

Ever since their characterization as a brand-new type of cell, the research field of TCs has continuously developed and improved. However, there are still numerous questions to be answered for them to be of use for doctors and patients alike. To achieve this goal, researchers should focus their efforts on discovering easier and more accessible ways of identifying their presence. Furthermore, their functions should be studied more in depth, by loss-of-function studies, both in vivo, through ablating them, and in vitro, through tissue engineering. Nevertheless, their research brings exciting perspectives and could be the key for a better understanding of the underlying mechanisms of certain pathologies and even lead to new and innovative therapies.

References

[1] Faussone Pellegrini MS, Popescu LM. Telocytes. Biomol Concept 2011;2(6):481–9.

[2] Bei Y, Wang F, Yang C, Xiao J. Telocytes in regenerative medicine. J Cell Mol Med 2015;19(7):1441–54.

[3] Popescu LM, Faussone-Pellegrini MS. TELOCYTES - a case of serendipity: the winding way from interstitial cells of Cajal (ICC), via interstitial Cajal-like cells (ICLC) to TELOCYTES. J Cell Mol Med 2010;14(4):729–40.

[4] Al-Shboul OA. The importance of interstitial cells of Cajal in the gastrointestinal tract. Saudi J Gastroenterol 2013;19(1):3–15.

[5] Thuneberg L. Interstitial cells of Cajal: intestinal pacemaker cells? Adv Anat Embryol Cell Biol 1982;71:1–130.

[6] Faussone Pellegrini MS, Cortesini C, Romagnoli P. [Ultrastructure of the tunica muscularis of the cardial portion of the human esophagus and stomach, with special reference to the so-called Cajal's interstitial cells]. Arch Ital Anat Embriol 1977;82(2): 157–77.

[7] Hinescu ME, Popescu LM. Interstitial Cajal-like cells (ICLC) in human atrial myocardium. J Cell Mol Med 2005;9(4):972–5.

[8] Gherghiceanu M, Popescu LM. Interstitial Cajal-like cells (ICLC) in human resting mammary gland stroma. Transmission electron microscope (TEM) identification. J Cell Mol Med 2005;9(4):893–910.

[9] Popescu LM, Gherghiceanu M, Cretoiu D, Radu E. The connective connection: interstitial cells of Cajal (ICC) and ICC-like cells establish synapses with immunoreactive cells. Electron microscope study in situ. J Cell Mol Med 2005;9(3):714–30.

[10] Popescu LM, Ciontea SM, Cretoiu D, Hinescu ME, Radu E, Ionescu N, et al. Novel type of interstitial cell (Cajal-like) in human fallopian tube. J Cell Mol Med 2005;9(2): 479–523.

[11] Popescu LM, Hinescu ME, Ionescu N, Ciontea SM, Cretoiu D, Ardelean C. Interstitial cells of Cajal in pancreas. J Cell Mol Med 2005;9(1):169–90.
[12] Wang J, Jin M, Ma WH, Zhu Z, Wang X. The history of telocyte discovery and understanding. Adv Exp Med Biol 2016;913:1–21.
[13] Pieri L, Vannucchi MG, Faussone-Pellegrini MS. Histochemical and ultrastructural characteristics of an interstitial cell type different from ICC and resident in the muscle coat of human gut. J Cell Mol Med 2008;12(5b):1944–55.
[14] Kostin S, Popescu LM. A distinct type of cell in myocardium: interstitial Cajal-like cells (ICLCs). J Cell Mol Med 2009;13(2):295–308.
[15] Robinson-Bennett B, Han A. 30 - role of immunohistochemistry in elucidating lung cancer metastatic to the ovary from primary ovarian carcinoma. In: Hayat MA, editor. Handbook of immunohistochemistry and in situ hybridization of human Carcinomas, vol. 4. Academic Press; 2006. p. 537–45.
[16] Varga I, Kyselovič J, Danišovič L, Gálfiová P, Kachlík D, Polák Š, et al. Recently discovered interstitial cells termed telocytes: distinguishing cell-biological and histological facts from fictions. Biologia 2019;74(2):195–203.
[17] Hatta K, Huang ML, Weisel RD, Li RK. Culture of rat endometrial telocytes. J Cell Mol Med 2012;16(7):1392–6.
[18] Yang XJ, Yang J, Liu Z, Yang G, Shen ZJ. Telocytes damage in endometriosis-affected rat oviduct and potential impact on fertility. J Cell Mol Med 2015;19(2): 452–62.
[19] Aleksandrovych V, Bialas M, Pasternak A, Bereza T, Sajewicz M, Walocha J, et al. Identification of uterine telocytes and their architecture in leiomyoma. Folia Med Cracov 2018;58(3):89–102.
[20] Varga I, Klein M, Urban L, Danihel Jr L, Polak S, Danihel Sr L. Recently discovered interstitial cells "telocytes" as players in the pathogenesis of uterine leiomyomas. Med Hypotheses 2018;110:64–7.
[21] Liu T, Wang S, Li Q, Huang Y, Chen C, Zheng J. Telocytes as potential targets in a cyclophosphamide-induced animal model of premature ovarian failure. Mol Med Rep 2016;14(3):2415–22.
[22] Yang P, Zhu X, Wang L, Ahmed N, Huang Y, Chen H, et al. Cellular evidence of telocytes as novel interstitial cells within the magnum of chicken oviduct. Cell Transplant 2017;26(1):135–43.
[23] Yang J, Chi C, Liu Z, Yang G, Shen ZJ, Yang XJ. Ultrastructure damage of oviduct telocytes in rat model of acute salpingitis. J Cell Mol Med 2015;19(7):1720–8.
[24] Aleksandrovych V, Sajewicz M, Walocha JA, Gil K. Tubal telocytes: factor infertility reason? Folia Med Cracov 2016;56(2):17–23.
[25] Petre N, Rusu MC, Pop F, Jianu AM. Telocytes of the mammary gland stroma. Folia Morphol 2016;75(2):224–31.
[26] Nizyaeva NV, Sukhacheva TV, Kulikova GV, Nagovitsyna MN, Poltavtseva RA, Kan NE, et al. Ultrastructural characteristics of placental telocytes. Bull Exp Biol Med 2017;162(5):693–8.
[27] Nizyaeva NV, Sukhacheva TV, Serov RA, Kulikova GV, Nagovitsyna MN, Kan NE, et al. Ultrastructural and immunohistochemical features of telocytes in placental villi in preeclampsia. Sci Rep 2018;8(1):3453.
[28] Marini M, Ibba-Manneschi L, Rosa I, Sgambati E, Manetti M. Changes in the telocyte/CD34+ stromal cell and alpha-SMA+ myoid cell networks in human testicular seminoma. Acta Histochem 2019;121(8):151442.
[29] Marini M, Rosa I, Guasti D, Gacci M, Sgambati E, Ibba-Manneschi L, et al. Reappraising the microscopic anatomy of human testis: identification of telocyte networks in the peritubular and intertubular stromal space. Sci Rep 2018;8(1):14780.

[30] Corradi LS, Jesus MM, Fochi RA, Vilamaior PS, Justulin Jr LA, Goes RM, et al. Structural and ultrastructural evidence for telocytes in prostate stroma. J Cell Mol Med 2013;17(3):398–406.
[31] Popescu LM, Manole CG, Gherghiceanu M, Ardelean A, Nicolescu MI, Hinescu ME, et al. Telocytes in human epicardium. J Cell Mol Med 2010;14(8): 2085–93.
[32] Shim W. Myocardial telocytes: a new player in electric circuitry of the heart. Adv Exp Med Biol 2016;913:241–51.
[33] Zhaofu L, Dongqing C. Cardiac telocytes in regeneration of myocardium after myocardial infarction. Adv Exp Med Biol 2016;913:229–39.
[34] Manole CG, Cismasiu V, Gherghiceanu M, Popescu LM. Experimental acute myocardial infarction: telocytes involvement in neo-angiogenesis. J Cell Mol Med 2011;15(11):2284–96.
[35] Gherghiceanu M, Manole CG, Popescu LM. Telocytes in endocardium: electron microscope evidence. J Cell Mol Med 2010;14(9):2330–4.
[36] Mitrofanova LB, Gorshkov AN, Konovalov PV, Krylova JS. Telocytes in the human sinoatrial node. J Cell Mol Med 2018;22(1):521–32.
[37] Hostiuc S, Negoi I, Dogaroiu C, Drima E, Iancu CB. Cardiac telocytes. From basic science to cardiac diseases. I. Atrial fibrillation. Ann Anat 2018;218:83–7.
[38] Yang Y, Sun W, Wu SM, Xiao J, Kong X. Telocytes in human heart valves. J Cell Mol Med 2014;18(5):759–65.
[39] Rosa I, Taverna C, Novelli L, Marini M, Ibba-Manneschi L, Manetti M. Telocytes constitute a widespread interstitial meshwork in the lamina propria and underlying striated muscle of human tongue. Sci Rep 2019;9(1):5858.
[40] Chen X, Zheng Y, Manole CG, Wang X, Wang Q. Telocytes in human oesophagus. J Cell Mol Med 2013;17(11):1506–12.
[41] Rusu MC, Nicolescu MI, Jianu AM, Lighezan R, Manoiu VS, Paduraru D. Esophageal telocytes and hybrid morphologies. Cell Biol Int 2012;36(12):1079–88.
[42] Zhang H, Zhong S, Yu P, Ge T, Peng S, Guo X, et al. Telocytes in gastric lamina propria of the Chinese giant salamander, *Andrias davidianus*. Sci Rep 2016;6:33554.
[43] Shoshkes-Carmel M, Wang YJ, Wangensteen KJ, Toth B, Kondo A, Massasa EE, et al. Subepithelial telocytes are an important source of Wnts that supports intestinal crypts. Nature 2018;557(7704):242–6.
[44] Manetti M, Rosa I, Messerini L, Ibba-Manneschi L. Telocytes are reduced during fibrotic remodelling of the colonic wall in ulcerative colitis. J Cell Mol Med 2015; 19(1):62–73.
[45] Ricci R, Giustiniani MC, Gessi M, Lanza P, Castri F, Biondi A, et al. Telocytes are the physiological counterpart of inflammatory fibroid polyps and PDGFRA-mutant GISTs. J Cell Mol Med 2018;22(10):4856–62.
[46] Chen L, Yu B. Telocytes and interstitial cells of Cajal in the biliary system. J Cell Mol Med 2018;22(7):3323–9.
[47] Pasternak A, Bugajska J, Szura M, Walocha JA, Matyja A, Gajda M, et al. Biliary polyunsaturated fatty acids and telocytes in gallstone disease. Cell Transplant 2017;26(1): 125–33.
[48] Zhang H, Yu P, Zhong S, Ge T, Peng S, Guo X, et al. Telocytes in pancreas of the Chinese giant salamander (*Andrias davidianus*). J Cell Mol Med 2016;20(11):2215–9.
[49] Liu J, Cao Y, Song Y, Huang Q, Wang F, Yang W, et al. Telocytes in liver. Curr Stem Cell Res Ther 2016;11(5):415–9.
[50] Zhao Y, Chen X, Wang F, Yang C. Hepatic telocytes. Adv Exp Med Biol 2016;913: 425–32.
[51] Fu S, Wang F, Cao Y, Huang Q, Xiao J, Yang C, et al. Telocytes in human liver fibrosis. J Cell Mol Med 2015;19(3):676–83.

[52] Rusu MC, Jianu AM, Mirancea N, Didilescu AC, Manoiu VS, Paduraru D. Tracheal telocytes. J Cell Mol Med 2012;16(2):401–5.
[53] Hinescu ME, Gherghiceanu M, Suciu L, Popescu LM. Telocytes in pleura: two- and three-dimensional imaging by transmission electron microscopy. Cell Tissue Res 2011;343(2):389–97.
[54] Zheng Y, Li H, Manole CG, Sun A, Ge J, Wang X. Telocytes in trachea and lungs. J Cell Mol Med 2011;15(10):2262–8.
[55] Popescu LM, Gherghiceanu M, Suciu LC, Manole CG, Hinescu ME. Telocytes and putative stem cells in the lungs: electron microscopy, electron tomography and laser scanning microscopy. Cell Tissue Res 2011;345(3):391–403.
[56] Alunno A, Ibba-Manneschi L, Bistoni O, Rosa I, Caterbi S, Gerli R, et al. Telocytes in minor salivary glands of primary Sjogren's syndrome: association with the extent of inflammation and ectopic lymphoid neogenesis. J Cell Mol Med 2015;19(7): 1689–96.
[57] Nicolescu MI, Bucur A, Dinca O, Rusu MC, Popescu LM. Telocytes in parotid glands. Anat Rec 2012;295(3):378–85.
[58] Qi G, Lin M, Xu M, Manole CG, Wang X, Zhu T. Telocytes in the human kidney cortex. J Cell Mol Med 2012;16(12):3116–22.
[59] Zheng L, Li L, Qi G, Hu M, Hu C, Wang S, et al. Transplantation of telocytes attenuates unilateral ureter obstruction-induced renal fibrosis in rats. Cell Physiol Biochem 2018;46(5):2056–71.
[60] Dobra MA, Vrapciu AD, Pop F, Petre N, Rusu MC. The molecular phenotypes of ureteral telocytes are layer-specific. Acta Histochem 2018;120(1):41–5.
[61] Vannucchi MG, Traini C. The telocytes/myofibroblasts 3-D network forms a stretch receptor in the human bladder mucosa. Is this structure involved in the detrusor overactive diseases? Ann Anat 2018;218:118–23.
[62] Traini C, Fausssone-Pellegrini MS, Guasti D, Del Popolo G, Frizzi J, Serni S, et al. Adaptive changes of telocytes in the urinary bladder of patients affected by neurogenic detrusor overactivity. J Cell Mol Med 2018;22(1):195–206.
[63] Rusu MC, Mirancea N, Manoiu VS, Valcu M, Nicolescu MI, Paduraru D. Skin telocytes. Ann Anat - Anat Anz 2012;194(4):359–67.
[64] Popescu LM, Manole E, Serboiu CS, Manole CG, Suciu LC, Gherghiceanu M, et al. Identification of telocytes in skeletal muscle interstitium: implication for muscle regeneration. J Cell Mol Med 2011;15(6):1379–92.
[65] Li H, Zhang H, Yang L, Lu S, Ge J. Telocytes in mice bone marrow: electron microscope evidence. J Cell Mol Med 2014;18(6):975–8.
[66] Zhang H. Vascular telocytes. Adv Exp Med Biol 2016;913:377–95.
[67] Rusu MC, Manoiu VS, Cretoiu D, Cretoiu SM, Vrapciu AD. Stromal cells/telocytes and endothelial progenitors in the perivascular niches of the trigeminal ganglion. Ann Anat 2018;218:141–55.
[68] Cantarero I, Luesma MJ, Alvarez-Dotu JM, Munoz E, Junquera C. Transmission electron microscopy as key technique for the characterization of telocytes. Curr Stem Cell Res Ther 2016;11(5):410–4.
[69] Cretoiu SM, Popescu LM. Telocytes revisited. Biomol Concept 2014;5(5):353–69.
[70] Cretoiu SM, Cretoiu D, Marin A, Radu BM, Popescu LM. Telocytes: ultrastructural, immunohistochemical and electrophysiological characteristics in human myometrium. Reproduction 2013;145(4):357–70.
[71] Cretoiu D. The third dimension of telocytes revealed by FIB-SEM tomography. Adv Exp Med Biol 2016;913:325–34.
[72] Cantarero I, Luesma MJ, Junquera C. The primary cilium of telocytes in the vasculature: electron microscope imaging. J Cell Mol Med 2011;15(12):2594–600.

[73] Cretoiu D, Xu J, Xiao J, Cretoiu SM. Telocytes and their extracellular vesicles-evidence and hypotheses. Int J Mol Sci 2016;17(8).
[74] Cretoiu SM, Simionescu AA, Caravia L, Curici A, Cretoiu D, Popescu LM. Complex effects of imatinib on spontaneous and oxytocin-induced contractions in human non-pregnant myometrium. Acta Physiologica Hungarica 2011;98(3):329−38.
[75] Rosenbaum ST, Svalo J, Nielsen K, Larsen T, Jorgensen JC, Bouchelouche P. Immunolocalization and expression of small-conductance calcium-activated potassium channels in human myometrium. J Cell Mol Med 2012;16(12):3001−8.
[76] Radu BM, Banciu A, Banciu DD, Radu M, Cretoiu D, Cretoiu SM. Calcium signaling in interstitial cells: focus on telocytes. Int J Mol Sci 2017;18(2).
[77] Zheng Y, Wang X. Roles of telocytes in the development of angiogenesis. Adv Exp Med Biol 2016;913:253−61.
[78] Vannucchi MG, Bani D, Faussone-Pellegrini MS. Telocytes contribute as cell progenitors and differentiation inductors in tissue regeneration. Curr Stem Cell Res Ther 2016;11(5):383−9.
[79] Jiang XJ, Cretoiu D, Shen ZJ, Yang XJ. An in vitro investigation of telocytes-educated macrophages: morphology, heterocellular junctions, apoptosis and invasion analysis. J Transl Med 2018;16(1):85.
[80] Hussein MM, Mokhtar DM. The roles of telocytes in lung development and angiogenesis: an immunohistochemical, ultrastructural, scanning electron microscopy and morphometrical study. Dev Biol 2018;443(2):137−52.
[81] Bani D. Telocytes in cardiac tissue architecture and development. Adv Exp Med Biol 2016;913:127−37.
[82] Popescu LM, Faussone-Pellegrini MS. Telocytes—a case of serendipity: the winding way from interstitial cells of Cajal (ICC), via interstitial Cajal-like cells (ICLC) to telocytes. J Cell Mol Med 2010;14(4):729−40.
[83] Faussone-Pellegrini MS, Gherghiceanu M. Telocyte's contacts. Semin Cell Dev Biol 2016;55:3−8.
[84] Gherghiceanu M, Popescu LM. Heterocellular communication in the heart: electron tomography of telocyte-myocyte junctions. J Cell Mol Med 2011;15:1005−11.
[85] Zheng YH, Li H, Ge JB, Gao HJ, Wang XD. [Location of telocytes in mouse bronchial and pulmonary tissues]. Zhonghua Bing Li Xue Za Zhi 2012;41:172−5.
[86] Cretoiu SM, Cretoiu D, Popescu LM. Human myometrium—the ultrastructural 3D network of telocytes. J Cell Mol Med 2012;16:2844−9.
[87] Camussi G, Deregibus MC, Bruno S, Cantaluppi V, Biancone L. Exosomes/microvesicles as a mechanism of cell-to-cell communication. Kidney Int 2010;78(9): 838−48.
[88] Corrado C, Raimondo S, Chiesi A, Ciccia F, De Leo G, Alessandro R. Exosomes as intercellular signaling organelles involved in health and disease: basic science and clinical applications. Int J Mol Sci 2013;14(3):5338−66.
[89] Emanueli C, Shearn AI, Angelini GD, Sahoo S. Exosomes and exosomal miRNAs in cardiovascular protection and repair. Vasc Pharmacol 2015;71:24−30.
[90] Vader P, Breakefield XO, Wood MJ. Extracellular vesicles: emerging targets for cancer therapy. Trends Mol Med 2014;20(7):385−93.
[91] Yoon YJ, Kim OY, Gho YS. Extracellular vesicles as emerging intercellular communicasomes. BMB Rep 2014;47(10):531−9.
[92] Ratajczak MZ, Ratajczak D, Pedziwiatr D. Extracellular microvesicles (ExMVs) in cell to cell communication: a role of telocytes. Adv Exp Med Biol 2016;913:41−9.
[93] EL Andaloussi S, Mäger I, Breakefield XO, Wood MJ. Extracellular vesicles: biology and emerging therapeutic opportunities. Nat Rev Drug Discov 2013;12(5):347−57.
[94] Cismaşiu VB, Popescu LM. Telocytes transfer extracellular vesicles loaded with microRNAs to stem cells. J Cell Mol Med 2015;19(2):351−8.

[95] Fertig ET, Gherghiceanu M, Popescu LM. Extracellular vesicles release by cardiac telocytes: electron microscopy and electron tomography. J Cell Mol Med 2015; 18(10):1938–43.
[96] Bani D, Formigli L, Gherghiceanu M, Faussone-Pellegrini MS. Telocytes as supporting cells for myocardial tissue organization in developing and adult heart. J Cell Mol Med 2010;14(10):2531–8.
[97] Bei Y, Zhou Q, Sun Q, Xiao J. Telocytes in cardiac regeneration and repair. Semin Cell Dev Biol 2016;55:14–21.
[98] Diaz-Flores L, Gutierrez R, Garcia MP, Sáez FJ, Díaz-Flores Jr L, Valladares F, Madrid JF. CD34+ stromal cells/fibroblasts/fibrocytes/telocytes as a tissue reserve and a principal source of mesenchymal cells. Location, morphology, function and role in pathology. Histol Histopathol 2014;29(7):831–70.
[99] Popescu LM, Gherghiceanu M, Manole CG, Faussone-Pellegrini MS. Cardiac renewing: interstitial Cajal-like cells nurse cardiomyocyte progenitors in epicardial stem cell niches. J Cell Mol Med 2009;13(5):866–86.
[100] Zhou J, Zhang Y, Wen X, Cao J, Li D, Lin Q, Wang H, Liu Z, Duan C, Wu K, Wang C. Telocytes accompanying cardiomyocyte in primary culture: two- and three-dimensional culture environment. J Cell Mol Med 2010;14(11):2641–5.
[101] Popescu LM, Curici A, Wang E, Zhang H, Hu S, Gherghiceanu M. Telocytes and putative stem cells in ageing human heart. J Cell Mol Med 2015;19(1):31–45.
[102] Sun X, Zheng M, Zhang M, Qian M, Zheng Y, Li M, Cretoiu D, Chen C, Chen L, Popescu LM, Wang X. Differences in the expression of chromosome 1 genes between lung telocytes and other cells: mesenchymal stem cells, fibroblasts, alveolar type II cells, airway epithelial cells and lymphocytes. J Cell Mol Med 2014;18(5): 801–10.
[103] Zheng Y, Zhang M, Qian M, Wang L, Cismasiu VB, Bai C, Popescu LM, Wang X. Genetic comparison of mouse lung telocytes with mesenchymal stem cells and fibroblasts. J Cell Mol Med 2013;17(4):567–77.
[104] Stem cells, their niches and the systemic environment: an aging network: hearing before the Genetics. 2008.
[105] Campeanu RA, Radu BM, Cretoiu SM, Banciu DD, Banciu A, Cretoiu D, et al. Near-infrared low-level laser stimulation of telocytes from human myometrium. Laser Med Sci 2014;29(6):1867–74.
[106] Ceafalan L, Gherghiceanu M, Popescu LM, Simionescu O. Telocytes in human skin–are they involved in skin regeneration? J Cell Mol Med 2012;16(7):1405–20.
[107] Li L, Lin M, Li L, Wang R, Zhang C, Qi G, et al. Renal telocytes contribute to the repair of ischemically injured renal tubules. J Cell Mol Med 2014;18(6):1144–56.
[108] Chaitow L. Telocytes: connective tissue repair and communication cells. J Bodyw Mov Ther 2017;21(2):231–3.
[109] Boos AM, Weigand A, Brodbeck R, Beier JP, Arkudas A, Horch RE. The potential role of telocytes in tissue engineering and regenerative medicine. Semin Cell Dev Biol 2016;55:70–8.
[110] Gherghiceanu M, Popescu LM. Cardiac telocytes - their junctions and functional implications. Cell Tissue Res 2012;348(2):265–79.
[111] Edelstein L, Smythies J. The role of telocytes in morphogenetic bioelectrical signaling: once more unto the breach. Front Mol Neurosci 2014;7.
[112] Fertig ET, Gherghiceanu M, Popescu LM. Extracellular vesicles release by cardiac telocytes: electron microscopy and electron tomography. J Cell Mol Med 2014; 18(10):1938–43.
[113] Albulescu R, Tanase C, Codrici E, Popescu DI, Cretoiu SM, Popescu LM. The secretome of myocardial telocytes modulates the activity of cardiac stem cells. J Cell Mol Med 2015;19(8):1783–94.

[114] Cretoiu D, Cretoiu SM. Telocytes in the reproductive organs: current understanding and future challenges. Semin Cell Dev Biol 2016;55:40–9.
[115] Milia AF, Ruffo M, Manetti M, Rosa I, Conte D, Fazi M, et al. Telocytes in Crohn's disease. J Cell Mol Med 2013;17(12):1525–36.
[116] Gherghiceanu M, Popescu LM. Cardiomyocyte precursors and telocytes in epicardial stem cell niche: electron microscope images. J Cell Mol Med 2010;14(4):871–7.
[117] Zheng Y, Cretoiu D, Yan G, Cretoiu SM, Popescu LM, Fang H, et al. Protein profiling of human lung telocytes and microvascular endothelial cells using iTRAQ quantitative proteomics. J Cell Mol Med 2014;18(6):1035–59.
[118] Wen Q, Tang EI, Gao Y, Jesus TT, Chu DS, Lee WM, et al. Signaling pathways regulating blood-tissue barriers - Lesson from the testis. Biochim Biophys Acta Biomembr 2018;1860(1):141–53.
[119] Xiao T, Guo W, Chen M, Hao C, Gao S, Huang J, et al. Fabrication and in vitro study of tissue-Engineered Cartilage scaffold derived from Wharton's Jelly extracellular matrix. 2017. p. 5839071.
[120] Smith SC, Baker PN, Symonds EM. Placental apoptosis in normal human pregnancy. Am J Obstet Gynecol 1997;177(1):57–65.
[121] Bosco C, Díaz E, Gutiérrez R, González J, Parra-Cordero M, Rodrigo R, et al. A putative role for telocytes in placental barrier impairment during preeclampsia. Med Hypotheses 2015;84(1):72–7.
[122] Graf R, Langer JU, Schonfelder G, et al. The extravascular contractile system in the human placenta. Morphological and immunocytochemical investigations. Anat Embryol 1994;190:541–8.
[123] Kohnen G, Kertschanska S, Demir R, Kaufmann P. Placental villous stroma as a model system for myofibroblast differentiation. Histochem Cell Biol 1996;105: 415–29.
[124] Suciu L, Popescu LM, Gherghiceanu M. Human placenta: de visu demonstration of interstitial Cajal-like cells. J Cell Mol Med 2007;11(3):590–7.
[125] Suciu L, Popescu LM, Gherghiceanu M, Regalia T, Nicolescu MI, Hinescu ME, et al. Telocytes in human term placenta: morphology and phenotype. Cells Tissues Organs 2010;192(5):325–39.
[126] Bosco C, et al. A putative role for telocytes in placental barrier impairment during preeclampsia. Med Hypotheses 2015;84 1:72–7.
[127] Suciu L, Popescu LM, Gherghiceanu M, et al. Telocytes in human term placenta: morphology and phenotype. Cells Tissues Organs 2010;192:325–39.
[128] Kurahashi M, Nakano Y, Peri LE, Townsend JB, Ward SM, Sanders KM. A novel population of subepithelial platelet-derived growth factor receptor alpha-positive cells in the mouse and human colon. Am J Physiol Gastrointest Liver Physiol 2013;304(9):G823–34.
[129] Cretoiu D, Cretoiu SM, Simionescu AA, Popescu LM. Telocytes, a distinct type of cell among the stromal cells present in the lamina propria of jejunum. Histol Histopathol 2012;27(8):1067–78.
[130] Turner JR. Intestinal mucosal barrier function in health and disease. Nat Rev Immunol 2009;9(11):799–809.
[131] Creamer B, Shorter RG, Bamforth J. The turnover and shedding of epithelial cells: I—the turnover in the gastro- intestinal tract. Gut 1961;2:110–8.
[132] Vannucchi MG, Traini C, Manetti M, Ibba-Manneschi L, Faussone-Pellegrini MS. Telocytes express PDGFRalpha in the human gastrointestinal tract. J Cell Mol Med 2013;17(9):1099–108.
[133] Cretoiu D, Vannucchi M, Bei Y, Manetti M, Faussone-Pellegrini M, Ibba-Manneschi L, et al. Telocytes: new connecting devices in the stromal space of organs. 2019.

[134] Ibba-Manneschi L, Rosa I, Manetti M. Telocyte implications in human pathology: an overview. Semin Cell Dev Biol 2016;55:62–9.
[135] Sanders KM, Ward SM, Koh SD. Interstitial cells: regulators of smooth muscle function. Physiol Rev 2014;94(3):859–907.
[136] Vannucchi MG, Traini C. Interstitial cells of Cajal and telocytes in the gut: twins, related or simply neighbor cells? Biomol Concept 2016;7(2):93–102.
[137] Shoshkes-Carmel M, Wang YJ, Wangensteen KJ, Toth B, Kondo A, Massasa EE, Itzkovitz S, Kaestner KH. Subepithelial telocytes are an important source of Wnts that supports intestinal crypts. Nature 2018;557:242–6.
[138] Aoki R, Shoshkes-Carmel M, Gao N, Shin S, May CL, Golson ML, Zahm AM, Ray M, Wiser CL, Wright CV, Kaestner KH. Foxl1-expressing mesenchymal cells constitute the intestinal stem cell niche. Cell Mol Gastroenterol Hepatol 2016;2: 175–88.
[139] Samuelson LC. Debate over the identity of an intestinal niche-cell population settled. Nature 2018;558:380–1.
[140] Elias PM. Stratum corneum defensive functions: an integrated view. J Invest Dermatol 2005;125(2):183–200.
[141] Del Rosso JQ, Levin J. The clinical relevance of maintaining the functional integrity of the stratum corneum in both healthy and disease-affected skin. J Clin Aesthet Dermatol 2011;4(9):22–42.
[142] McLafferty E, Hendry C, Alistair F. The integumentary system: anatomy, physiology and function of skin. Nurs Stand 2012;27(3):35–42.
[143] Nafisi S, Maibach HI. Chapter 3 - skin penetration of nanoparticles. In: Shegokar R, Souto EB, editors. Emerging nanotechnologies in immunology. Boston: Elsevier; 2018. p. 47–88.
[144] Manole CG, Simionescu O. The cutaneous telocytes. Adv Exp Med Biol 2016;913: 303–23.
[145] Cretoiu D, Gherghiceanu M, Hummel E, Zimmermann H, Simionescu O, Popescu LM. FIB-SEM tomography of human skin telocytes and their extracellular vesicles. J Cell Mol Med 2015;19(4):714–22.
[146] Manetti M, Guiducci S, Ruffo M, Rosa I, Faussone-Pellegrini MS, Matucci-Cerinic M, et al. Evidence for progressive reduction and loss of telocytes in the dermal cellular network of systemic sclerosis. J Cell Mol Med 2013;17(4):482–96.
[147] Manole CG, Gherghiceanu M, Simionescu O. Telocyte dynamics in psoriasis. J Cell Mol Med 2015;19(7):1504–19.
[148] Zheng Y, Cretoiu D, Yan G, Cretoiu SM, Popescu LM, Wang X. Comparative proteomic analysis of human lung telocytes with fibroblasts. J Cell Mol Med 2014;18(4): 568–89.
[149] Niculite CM, Regalia TM, Gherghiceanu M, Huica R, Surcel M, Ursaciuc C, et al. Dynamics of telopodes (telocyte prolongations) in cell culture depends on extracellular matrix protein. Mol Cell Biochem 2015;398(1–2):157–64.
[150] Richter M, Kostin S. The failing human heart is characterized by decreased numbers of telocytes as result of apoptosis and altered extracellular matrix composition. J Cell Mol Med 2015;19(11):2597–606.
[151] Naylor AJ, Filer A, Buckley CD. The role of stromal cells in the persistence of chronic inflammation. Clin Exp Immunol 2013;171(1):30–5.
[152] Varga I, Urban L, Kajanova M, Polak S. Functional histology and possible clinical significance of recently discovered telocytes inside the female reproductive system. Arch Gynecol Obstet 2016;294(2):417–22.
[153] Mihalcea CE, Morosanu AM, Murarasu D, Puiu L, Cinca SA, Voinea SC, et al. Molecular analysis of BRCA1 and BRCA2 genes by next generation sequencing and

ultrastructural aspects of breast tumor tissue. Rom J Morphol Embryol 2017;58(2): 445–55.
[154] Ellison G, Huang S, Carr H, Wallace A, Ahdesmaki M, Bhaskar S, et al. A reliable method for the detection of BRCA1 and BRCA2 mutations in fixed tumour tissue utilising multiplex PCR-based targeted next generation sequencing. BMC Clin Pathol 2015;15:5.
[155] Muschler J, Streuli CH. Cell-matrix interactions in mammary gland development and breast cancer. Cold Spring Harb Perspect Biol 2010;2(10):a003202.
[156] Mou Y, Wang Y, Li J, Lü S, Duan C, Du Z, et al. Immunohistochemical characterization and functional identification of mammary gland telocytes in the self-assembly of reconstituted breast cancer tissue in vitro. J Cell Mol Med 2013;17(1):65–75.
[157] Cretoiu SM. Immunohistochemistry of telocytes in the uterus and fallopian tubes. Adv Exp Med Biol 2016;913:335–57.
[158] Ciontea SM, Radu E, Regalia T, Ceafalan L, Cretoiu D, Gherghiceanu M, et al. C-kit immunopositive interstitial cells (Cajal-type) in human myometrium. J Cell Mol Med 2005;9(2):407–20.
[159] Xiao J, Wang F, Liu Z, Yang C. Telocytes in liver: electron microscopic and immunofluorescent evidence. J Cell Mol Med 2013;17(12):1537–42.
[160] Cretoiu D, Ciontea SM, Popescu LM, Ceafalan L, Ardeleanu C. Interstitial Cajal-like cells (ICLC) as steroid hormone sensors in human myometrium: immunocytochemical approach. J Cell Mol Med 2006;10(3):789–95.
[161] Hinescu ME, Popescu LM, Gherghiceanu M, Faussone-Pellegrini MS. Interstitial Cajal-like cells in rat mesentery: an ultrastructural and immunohistochemical approach. J Cell Mol Med 2008;12(1):260–70.
[162] Evans WH, Martin PE. Gap junctions: structure and function (review). Mol Membr Biol 2002;19(2):121–36.
[163] Laws MJ, Taylor RN, Sidell N, DeMayo FJ, Lydon JP, Gutstein DE, et al. Gap junction communication between uterine stromal cells plays a critical role in pregnancy-associated neovascularization and embryo survival. Development 2008;135(15): 2659–68.
[164] Nair RR, Jain M, Singh K. Reduced expression of gap junction gene connexin 43 in recurrent early pregnancy loss patients. Placenta 2011;32(8):619–21.
[165] Bastiani M, Parton RG. Caveolae at a glance. J Cell Sci 2010;123(Pt 22):3831–6.
[166] Paulin D, Li Z. Desmin: a major intermediate filament protein essential for the structural integrity and function of muscle. Exp Cell Res 2004;301(1):1–7.
[167] Shinde AV, Humeres C, Frangogiannis NG. The role of α-smooth muscle actin in fibroblast-mediated matrix contraction and remodeling. Biochim Biophys Acta - Mol Basis Dis 2017;1863(1):298–309.
[168] Aumailley M. The laminin family. Cell Adh Migr 2013;7(1):48–55.
[169] Abd-Elhafeez HH, Soliman SA. New description of telocyte sheaths in the bovine uterine tube: an immunohistochemical and scanning microscopic study. Cells Tissues Organs 2017;203(5):295–315.
[170] Janke C. The tubulin code: molecular components, readout mechanisms, and functions. J Cell Biol 2014;206(4):461–72.
[171] Gandahi JA, Chen SF, Yang P, Bian XG, Chen QS. Ultrastructural identification of interstitial cells of Cajal in hen oviduct. Poultry Sci 2012;91(6):1410–7.
[172] Cantarero Carmona I, Luesma Bartolome MJ, Junquera Escribano C. Identification of telocytes in the lamina propria of rat duodenum: transmission electron microscopy. J Cell Mol Med 2011;15(1):26–30.
[173] Gevaert T, De Vos R, Van Der Aa F, Joniau S, van den Oord J, Roskams T, et al. Identification of telocytes in the upper lamina propria of the human urinary tract. J Cell Mol Med 2012;16(9):2085–93.

[174] Roatesi I, Radu BM, Cretoiu D, Cretoiu SM. Uterine telocytes: a review of current knowledge. Biol Reprod 2015;93(1):10.

[175] Duquette RA, Shmygol A, Vaillant C, Mobasheri A, Pope M, Burdyga T, et al. Vimentin-positive, c-KIT-negative interstitial cells in human and rat uterus: a role in pacemaking? Biol Reprod 2005;72(2):276−83.

[176] Salama N. Immunohistochemical characterization of telocytes in ratuterus in different reproductive states. 2013. p. 185−94.

[177] Krall JF, Swensen JL, Korenman SG. Hormonal control of uterine contraction. Characterization od cyclic AMP-dependent membrane properties in the myometrium. Biochim Biophys Acta 1976;448(4):578−88.

[178] Hutchings G, Williams O, Cretoiu D, Ciontea SM. Myometrial interstitial cells and the coordination of myometrial contractility. J Cell Mol Med 2009;13(10):4268−82.

[179] Horn LC, Meinel A, Hentschel B. c-kit/CD 117 positive cells in the myometrium of pregnant women and those with uterine endometriosis. Arch Gynecol Obstet 2012; 286(1):105−7.

[180] Chi C, Jiang XJ, Su L, Shen ZJ, Yang XJ. In vitro morphology, viability and cytokine secretion of uterine telocyte-activated mouse peritoneal macrophages. J Cell Mol Med 2015;19(12):2741−50.

[181] Yang XJ, Xu JY, Shen ZJ, Zhao J. Immunohistochemical alterations of Cajal-like type of tubal interstitial cells in women with endometriosis and tubal ectopic pregnancy. Arch Gynecol Obstet 2013;288(6):1295−300.

[182] Kostin S. Myocardial telocytes: a specific new cellular entity. J Cell Mol Med 2010; 14(7):1917−21.

[183] Faussone-Pellegrini M-S, Bani D. Relationships between telocytes and cardiomyocytes during pre- and post-natal life. J Cell Mol Med 2010;14(5):1061−3.

[184] Zhou J, Wang Y, Zhu P, Sun H, Mou Y, Duan C, et al. Distribution and characteristics of telocytes as nurse cells in the architectural organization of engineered heart tissues. Sci China Life Sci 2014;57(2):241−7.

[185] Bei Y, Zhou Q, Fu S, Lv D, Chen P, Chen Y, et al. Cardiac telocytes and fibroblasts in primary culture: different morphologies and immunophenotypes. PLoS One 2015; 10(2):e0115991.

[186] Diaz-Flores L, Gutierrez R, Madrid JF, Varela H, Valladares F, Acosta E, et al. Pericytes. Morphofunction, interactions and pathology in a quiescent and activated mesenchymal cell niche. Histol Histopathol 2009;24(7):909−69.

[187] Popescu LM, Gherghiceanu M, Hinescu ME, Cretoiu D, Ceafalan L, Regalia T, et al. Insights into the interstitium of ventricular myocardium: interstitial Cajal-like cells (ICLC). J Cell Mol Med 2006;10(2):429−58.

[188] Rusu MC, Pop F, Hostiuc S, Curca GC, Jianu AM, Paduraru D. Telocytes form networks in normal cardiac tissues. Histol Histopathol 2012;27(6):807−16.

[189] Zhou Q, Wei L, Zhong C, Fu S, Bei Y, Huică R-I, et al. Cardiac telocytes are double positive for CD34/PDGFR-α. J Cell Mol Med 2015;19(8):2036−42.

[190] Kostin S. Cardiac telocytes in normal and diseased hearts. Semin Cell Dev Biol 2016; 55:22−30.

[191] Barile L, Gherghiceanu M, Popescu LM, Moccetti T, Vassalli G. Ultrastructural evidence of exosome secretion by progenitor cells in adult mouse myocardium and adult human cardiospheres. J Biomed Biotechnol 2012;2012:354605.

[192] Smythies J, Edelstein L. Telocytes, exosomes, gap junctions and the cytoskeleton: the makings of a primitive nervous system? Front Cell Neurosci 2014;7:278.

[193] Zhao B, Chen S, Liu J, Yuan Z, Qi X, Qin J, et al. Cardiac telocytes were decreased during myocardial infarction and their therapeutic effects for ischaemic heart in rat. J Cell Mol Med 2013;17(1):123−33.

[194] Manetti M, Rosa I, Messerini L, Guiducci S, Matucci-Cerinic M, Ibba-Manneschi L. A loss of telocytes accompanies fibrosis of multiple organs in systemic sclerosis. J Cell Mol Med 2014;18(2):253–62.

[195] Enciu AM, Popescu LM. Telopodes of telocytes are influenced in vitro by redox conditions and ageing. Mol Cell Biochem 2015;410(1–2):165–74.

[196] Zheng Y, Chen X, Qian M, Zhang M, Zhang D, Bai C, et al. Human lung telocytes could promote the proliferation and angiogenesis of human pulmonary microvascular endothelial cells in vitro. Mol Cell Ther 2014;2:3.

[197] Zheng Y, Bai C, Wang X. Potential significance of telocytes in the pathogenesis of lung diseases. Expert Rev Respir Med 2012;6(1):45–9.

[198] Wang X-Y, Berezin I, Mikkelsen HB, Der T, Bercik P, Collins SM, et al. Pathology of interstitial cells of Cajal in relation to inflammation revealed by ultrastructure but not immunohistochemistry. Am J Pathol 2002;160(4):1529–40.

[199] Popescu LM, Andrei F, Hinescu ME. Snapshots of mammary gland interstitial cells: methylene-blue vital staining and c-kit immunopositivity. J Cell Mol Med 2005;9(2):476–7.

[200] Vijayan VK. Chronic obstructive pulmonary disease. Indian J Med Res 2013;137(2):251–69.

[201] Zheng Y, Bai C, Wang X. Telocyte morphologies and potential roles in diseases. J Cell Physiol 2012;227(6):2311–7.

[202] King Jr TE. Clinical advances in the diagnosis and therapy of the interstitial lung diseases. Am J Respir Crit Care Med 2005;172(3):268–79.

[203] Pellegrini Maria-Simonetta F, Popescu Laurențiu M. Telocytes. BioMol Concept 2011:481.

[204] Gibbons SJ, De Giorgio R, Faussone Pellegrini MS, Garrity-Park MM, Miller SM, Schmalz PF, et al. Apoptotic cell death of human interstitial cells of Cajal. Neuro Gastroenterol Motil 2009;21(1):85–93.

[205] Huizinga JD, White EJ. Progenitor cells of interstitial cells of Cajal: on the road to tissue repair. Gastroenterology 2008;134(4):1252–4.

[206] Pasternak A, Gil K, Matyja A, Gajda M, Sztefko K, Walocha JA, et al. Loss of gallbladder interstitial Cajal-like cells in patients with cholelithiasis. Neuro Gastroenterol Motil 2013;25(1):e17–24.

[207] Pasternak A, Matyja A, Gil K, Gajda M, Tomaszewski KA, Gajda M, et al. Interstitial cajal-like cells and bile lithogenicity in the pathogenesis of gall-stone disease. Pol Przegl Chir 2013;85(6):311–6.

[208] Portincasa P, Moschetta A, Palasciano G. Cholesterol gallstone disease. Lancet 2006; 368(9531):230–9.

[209] van Erpecum KJ. Biliary lipids, water and cholesterol gallstones. Biol Cell 2005; 97(11):815–22.

[210] Venneman NG, van Erpecum KJ. Pathogenesis of gallstones. Gastroenterol Clin North Am 2010;39(2):171–83 [vii].

[211] Apstein MD, Carey MC. Pathogenesis of cholesterol gallstones: a parsimonious hypothesis. Eur J Clin Invest 1996;26(5):343–52.

[212] Shaffer EA. The role of the gallbladder in gallstone formation. In: Fisher MM, Goresky CA, Shaffer EA, Strasberg SM, editors. Gallstones. Boston, MA: Springer US; 1979. p. 223–49.

[213] Ortiz-Hidalgo C, De Leon Bojorge B, Albores-Saavedra J. Stromal tumor of the gallbladder with phenotype of interstitial cells of Cajal: a previously unrecognized neoplasm. Am J Surg Pathol 2000;24(10):1420–3.

[214] Mendoza-Marin M, Hoang MP, Albores-Saavedra J. Malignant stromal tumor of the gallbladder with interstitial cells of Cajal phenotype. Arch Pathol Lab Med 2002; 126(4):481–3.

[215] Lavoie B, Balemba OB, Nelson MT, Ward SM, Mawe GM. Morphological and physiological evidence for interstitial cell of Cajal-like cells in the Guinea pig gallbladder. J Physiol 2007;579(Pt 2):487–501.
[216] Balemba OB, Bartoo AC, Nelson MT, Mawe GM. Role of mitochondria in spontaneous rhythmic activity and intracellular calcium waves in the Guinea pig gallbladder smooth muscle. Am J Physiol Gastrointest Liver Physiol 2008;294(2): G467–76.
[217] Huang Y, Mei F, Yu B, Zhang HJ, Han J, Jiang ZY, et al. Distribution of the interstitial Cajal-like cells in the gallbladder and extrahepatic biliary duct of the Guinea-pig. Acta Histochem 2009;111(2):157–65.
[218] Xu D, Yu BP, Luo HS, Chen LD. Control of gallbladder contractions by cholecystokinin through cholecystokinin-A receptors on gallbladder interstitial cells of Cajal. World J Gastroenterol 2008;14(18):2882–7.
[219] Hinescu ME, Ardeleanu C, Gherghiceanu M, Popescu LM. Interstitial Cajal-like cells in human gallbladder. J Mol Histol 2007;38(4):275–84.
[220] Popescu L, Nicolescu M. Telocytes and stem cells. 2013. p. 205–31.
[221] Hernandez-Gea V, Friedman SL. Pathogenesis of liver fibrosis. Annu Rev Pathol 2011;6:425–56.
[222] Bonder A, Tapper EB, Afdhal NH. Contemporary assessment of hepatic fibrosis. Clin Liver Dis 2015;19(1):123–34.
[223] Iwaisako K, Taura K, Koyama Y, Takemoto K, Asagiri M. Strategies to detect hepatic myofibroblasts in liver cirrhosis of different etiologies. Curr Pathobiol Rep 2014;2(4): 209–15.
[224] Li LC, Gao J, Li J. Emerging role of HMGB1 in fibrotic diseases. J Cell Mol Med 2014;18(12):2331–9.
[225] Grieb G, Bucala R. Fibrocytes in fibrotic diseases and wound healing. Adv Wound Care 2012;1(1):36–40.
[226] Wynn TA, Ramalingam TR. Mechanisms of fibrosis: therapeutic translation for fibrotic disease. Nat Med 2012;18(7):1028–40.
[227] Fausther M, Lavoie EG, Dranoff JA. Contribution of myofibroblasts of different origins to liver fibrosis. Curr Pathobiol Rep 2013;1(3):225–30.
[228] Iwaisako K, Jiang C, Zhang M, Cong M, Moore-Morris TJ, Park TJ, et al. Origin of myofibroblasts in the fibrotic liver in mice. Proc Natl Acad Sci USA 2014;111(32): E3297–305.
[229] Wang F, Song Y, Bei Y, Zhao Y, Xiao J, Yang C. Telocytes in liver regeneration: possible roles. J Cell Mol Med 2014;18(9):1720–6.
[230] Maul J, Zeitz M. Ulcerative colitis: immune function, tissue fibrosis and current therapeutic considerations. Langenbeck's Arch Surg 2012;397(1):1–10.
[231] Rieder F, Fiocchi C. Intestinal fibrosis in inflammatory bowel disease - current knowledge and future perspectives. J Crohns Colitis 2008;2(4):279–90.
[232] Zhao B, Liao Z, Chen S, Yuan Z, Yilin C, Lee KK, et al. Intramyocardial transplantation of cardiac telocytes decreases myocardial infarction and improves post-infarcted cardiac function in rats. J Cell Mol Med 2014;18(5):780–9.
[233] Ja KP, Miao Q, Zhen Tee NG, Lim SY, Nandihalli M, Ramachandra CJA, et al. iPSC-derived human cardiac progenitor cells improve ventricular remodelling via angiogenesis and interstitial networking of infarcted myocardium. J Cell Mol Med 2016;20(2):323–32.
[234] Popescu BO, Gherghiceanu M, Kostin S, Ceafalan L, Popescu LM. Telocytes in meninges and choroid plexus. Neurosci Lett 2012;516(2):265–9.

CHAPTER TWO

Effects of radiation on endothelial barrier and vascular integrity

Roxane M. Bouten[1], Erik F. Young[2], Reed Selwyn[3], Diego Iacono[4,5,6] W. Bradley Rittase[1], Regina M. Day[1]

[1]Department of Pharmacology and Molecular Therapeutics, Uniformed Services University of the Health Sciences, Bethesda, MD, United States
[2]Department of Electrical Engineering, Columbia University, New York, NY, United States
[3]Department of Radiology, University of New Mexico, Albuquerque, NM, United States
[4]Departments of Neurology, Pathology, and Anatomy, Physiology and Genetics, Uniformed Services University of the Health Sciences, Bethesda, MD, United States
[5]Neurodegenerative Clinic, National Institute of Neurological Disorders and Stroke (NINDS), NIH, Bethesda, MD, United States
[6]Department of Anatomy, Physiology, and Genetics, Uniformed Services University of the Health Sciences, Bethesda, MD, United States

1. Introduction: radiation-induced permeability of the vasculature

Following the discovery of X-rays and ionizing radiation in late 1890's and early 1900's by Wilhelm Conrad Röntgen, Antoine Becquerel, Marie Curie, and Pierre Curie, beneficial applications for these formidable energy sources were rigorously investigated [1–3]. However, by 1896, several reports were made of injuries from radiation exposure [4,5]. By the 1920's, multiple researchers, physicians, and engineers experimenting with the potential uses of radiation suffered from the hazards of their work, reporting skin burns, hair loss, and deep tissue damage, with a number of investigators developing fatal cancers [4,6,7]. Likewise, many of the earliest patients, for whom radiation was being used for imaging and cancer treatment, also exhibited radiation injuries [4,6]. It was quickly understood that, in learning to harness the power of radiation, great care had to be taken to avoid its adverse effects.

Between the 1890's and 1940's, early studies suggested that radiation effects on vascular tissue were pathologically significant injuries in humans [5,8,9]. Initial insight into radiation vascular injuries was obtained from gross pathological and histological examinations using tissues from patients who received experimental radiation therapy, from researchers with accidental radiation injury, and, rather surprisingly, from studies of irradiated healthy

Tissue Barriers in Disease, Injury and Regeneration
ISBN: 978-0-12-818561-2
https://doi.org/10.1016/B978-0-12-818561-2.00007-2

human volunteers [7,8,10–16]. Gassmann detailed endothelial cell swelling near radiation-induced skin ulcerations as an early cellular effect of radiation [9,11]. Rapid damage to capillaries was characterized by "excessive dilation" with hemorrhage, red and white blood cell aggregation, erythema, edema, blister formation, and in some cases, vascular congestion [11,17]. Baermann and Lisner, Unna, and Lisner also described delayed gross pathology and histological alterations of blood vessel: leukocyte infiltration, dilation of vessels, and swelling of the endothelium [8,10,11,18]. Based on these findings, endothelial cells were hypothesized to be "highly radiosensitive," and it was thought that capillaries were especially sensitive to radiation compared with larger vessels [13,19].

As more experience with the biological effects of ionizing radiation accumulated, it was recognized that radiation-induced vascular changes were sustained, and even progressive, from months to years after initial exposure, referred to as "late effects" [20,21]. In some cases, radiation caused a persistent loss of the vascular endothelium, often with increased inflammation in vessel adventitia and thickening of the intima, with alterations in the vascular smooth muscle, and fibrous plaques that could lead to vessel occlusion [8,10,11,18,22]. In tissues recovering from radiation effects, capillaries were dilated with altered, circuitous paths, and telangiectasia [11,22]. It was also noted that, following radiation exposure, vascular beds had altered responses to endogenous vasodilators; this was shown to involve a reduced capacity of the endothelium for the production of nitric oxide for smooth muscle relaxation [23,24]. Vascular alterations were believed, in some cases, to contribute to late radiation effects in surrounding tissues, such as inflammatory edema, ischemia, and necrosis [11,13,22]. Long-term human studies provided evidence that larger vessels were also affected by radiation with delayed pathological effects [25–29]. For example, epidemiological studies of atomic bomb survivors showed elevation of cardiovascular diseases, with a 9% increase in stroke and a 14% increase in coronary events [11,25,28,30,31].

Research in the 1940's and 50's suggested that multiple mechanisms were responsible for increased endothelial barrier permeability following radiation exposure [32]. It is now accepted that alterations to vascular permeability by radiation occur via direct and indirect mechanisms. Direct mechanisms from radiation include the rapid deposition of energy in the tissues, causing direct damage to biological macromolecules and generating reactive oxygen species. This sudden macromolecular damage induces redox stress, single- and double-stranded DNA breaks, and oxidation of other biological molecules to initiate signal transduction cascades that alter the barrier function and survival status of the cells of the endothelial layer [33]. Indirect mechanisms of radiation effects on the endothelial barrier have also been described, including activation of the immune system and the release of iron, each which effect the endothelium independently from immediate impact of radiation. The combination of direct and indirect radiation effects on the vascular function contribute to delayed and late damage in other tissues, including constitutive low-level inflammation, tissue necrosis and fibrosis, and permanent alterations in tissue architecture [21,29,31,33–37].

2. Direct mechanisms of radiation-induced vascular effects

The exposure of biological tissues to ionizing radiation results in the rapid deposition of high amounts of energy with the capacity to induce damage at the molecular level [38]. Energy absorption occurs within 10^{-17} to 10^{-13} s of radiation exposure to living cells [39]. Ionization of hydroxyl radicals and secondary radicals occurs within 10^{-10} to 10^{-6} s, and DNA radicals can be detected within 10^{-6} to 10^{-3} s [39]. The oxidation of and damage to DNA can be detected within seconds to hours following ionizing radiation exposure [39]. Although DNA is often considered to be a critical target of radiation damage [38], other macromolecules, including proteins, lipids, and other metabolites, undergo modification, such as oxidization and nitrosylation, which can lead to destabilization of basic cellular functions [31,40–46]. Radiation-induced modification of biological molecules generates a variety of cellular responses depending upon the degree of oxidative stress and modifications that are acquired [31]. Cellular reactions can range from the activation of specific signal transduction, in an effort to regain homeostasis following oxidative and nitrosative stress, to apoptosis, senescence, or necrotic cell death, when the cell is confronted by increasing levels of irreversible damage [21,31,41].

2.1 Radiation-induced effects on the endothelial barrier

The continuity of the endothelial barrier is critical for normal tissue homeostasis [47,48]. Regulated passage of gases, nutrients, water, ions, and macromolecules require the dynamic functions of the endothelial barrier [47]. Vascular endothelial cells are integrally involved in the maintenance of vascular permeability through the control of their cell–cell contacts and intracellular transport for transcytosis [47–51]. Intercellular endothelial transport, also called paracellular transport, is affected by the regulation of proteins directly involved in cell–cell junctions as well as by intracellular proteins, such as the cytoskeleton which can control cell connections through cellular contraction [52–54]. Intracellular molecular transport mediates the movement of molecules from the blood to the interstitium *through* the endothelial cell (also called transcytosis), using channels, transporters, or vesicles [51,55–60]. Optimal endothelial barrier function requires maintenance of both types of transport.

Studies have provided evidence that radiation exposure destabilizes the endothelial barrier, increasing vascular permeability and altering vesicular trafficking. These effects have been studied in vivo in a wide variety of animal models [17,19,61–64] and in vitro using cultured primary or immortalized endothelial cells [35,65,66]. In vivo studies have allowed the determination of the time course and doses of radiation that give rise to vascular damage, while in vitro studies provided valuable and complimentary information for the elucidation of molecular signaling pathways leading to the disruption/alteration of the endothelial barrier [48,66].

2.1.1 In vivo models of radiation-induced vascular permeability

As noted above, initial reports of radiation effects on the vasculature were made based on gross pathological and histological observations of the dermal vasculature from radiation researchers (with accidental injuries) and early patients exposed to clinical radiation for imaging and cancer treatment [14,31]. Erythema, observed as an early response to radiation, was interpreted as response to changes in vascular permeability, and histological samples provided evidence of alterations in capillaries and larger vessels within the irradiation field [14]. Capillaries were characterized as excessively dilated (telangiectasia), with hemorrhage and blood cell aggregation, which were thought to be causative factors for erythema, edema, blister formation, and vascular congestion [11,17,22].

Initial animal studies of radiation vascular effects utilized the release of colored dyes (e.g., pontamine blue or Evans blue dye) injected intravenously to detect vascular leak by gross pathological examination [67]. Jolles and Harrison reported the "blue flush effect" in the skin of rabbits, in which pontamine blue dye was released within 2 h following exposure to ionizing radiation (800 roentgen [R]) [68]. The area of dye release was well defined and could be formed into a grid pattern using a patterned lead shield pressed against the skin, suggesting capillary permeabilization was localized [68].

Radiation-induced changes in vascular permeability were subsequently demonstrated in blood vessels of the brain of rats, using Evans blue dye to show increased permeability at 18 h following exposure to 15 Gy irradiation [69]. These findings led to the hypothesis that radiation-induced permeability was a consequence in all vascular beds. Studies of radiation-induced dye release provided evidence that capillary permeability had multiple phases, with an initial weak, transient phase within minutes of radiation exposure (1000 R), followed by a second stronger phase that occurred ~30 min after irradiation and peaked at ~90 min postirradiation, called the "delayed phase" [70]. A third phase of sustained permeability was also observed, with varying intensity over 9–16 days postirradiation [68]. Interestingly, treatments with protease inhibitors blocked the delayed response with little or no effect on the immediate response [70]. These findings suggested that there were independent phases of vascular permeability, probably with independent mechanisms [70,71].

Electron microscopy was used to identify radiation-induced alterations at the cellular level in capillaries of cardiac microvasculature from rabbits [36]. Following exposure of the heart to 10–13 Gy, researchers observed "increased pinocytic transport across the endothelium, as demonstrated by ferritin distribution" as well as evidence of "intercellular junctional gaps or disruption of endothelial sheets, as indicated by leakage of carbon particles" [36,72]. Histological evidence suggested that endothelial cell phagocytosis and erythrophagocytosis were elevated following irradiation and were radiation dose-related [36,61,73]. These were the first data to provide direct evidence for both increased intracellular traffic and decreased intercellular connectivity that could lead to increased barrier permeability following radiation exposure.

In vivo studies of the molecular effects of radiation exposure were conducted using the bone marrow vasculature in a murine model of radiation hematopoietic injury [74]. Microarrays of gene expression in mice exposed to 5 Gy (0.6 Gy/min) total body irradiation at 6 h, 24 h, and 14 days postirradiation were compared with gene expression from nonirradiated control animals. Data showed that radiation-induced altered expression of some solute carriers: solute carrier (Slc) 22a14, Slc4a1, Slc7a11, Slc30a10, and Slc16a10, Ankyrin 2 (a protein which plays a role in endocytosis and intracellular transport), and some cell adhesion molecules (close homolog of L1 [Chl1] and platelet endothelial cell adhesion molecule) [74]. Many of the changes in gene expression were transient, and initial recovery of the bone marrow endothelium from the acute effects of radiation was observed ~14 days postirradiation. Although this study did not investigate effects on the endothelial barrier function, these changes are suggestive of alterations in the movement of solutes through endothelial cells and in connectivity between endothelial cells during the period before 14 days postirradiation. Late (>1 month postirradiation) alterations in endothelial gene expression, which could be indicative of permanent alterations in barrier function, have not yet been investigated [74].

The effects of high-energy particulate radiation on the vasculature have been compared with the effects of electromagnetic X-rays and gamma rays. Uncharged particles, such as neutrons, and charged particles, such as alpha particles or heavy charged particles, have higher linear energy transfer (LET) than either X-ray or gamma radiation, and, in general, charged particles cause more biomolecular damage per distance traveled. For instance, particulate radiation induces higher numbers of DNA radicals for distance traveled compared with X-ray or gamma radiation [39]. The capacity of charged particles to damage vascular tissue was initially reported based on observations of patients treated with radium and in workers who used radium paint [75–77]. Similar to what was observed for gamma ionizing radiation, charged particle injuries to the vasculature were characterized by fibrosis and the induction of altered blood flow, with circuitous pathways induced in small vessels [75]. Animal model studies performed at the Lawrence Berkeley Laboratory and studies of human patients who received radiosurgery for brain tumors provided additional evidence of changes in the microvasculature following particle radiation exposure [75,78,79]. In these studies, alterations to the blood brain barrier (BBB) were examined using fluorodeoxyglucose positron emission tomography imaging, gadolinium diethylene-triamine-pentaacetate enhanced magnetic resonance imaging (MRI) for evaluating permeability changes, and rubidium PET imaging [79–82]. Helium ions and other charge particles induced late alterations in microvasculature leading to vasogenic edema, suppression of metabolism, and parenchymal necrosis [75,78,79].

Together, the in vivo studies established both the time course and dose response of vascular effects of radiation exposure. Similar observations were made for the effects of radiation on a variety of vascular beds, suggesting that the radiation response of endothelial cells may be similar in different tissues. Likewise, comparisons of different types of radiation suggest that there is consistency in the effects of radiation on vasculature in response, toward increased vascular permeability, to different energies. These studies also provided initial insights into gene expression and protein changes that occurred over time.

2.1.2 *In vitro signal transduction by ionizing radiation for endothelial barrier permeability*

In vitro studies of barrier function complement in vivo studies and allow for the identification of molecular mechanisms and signaling pathways leading to changes in barrier permeability. The techniques for studying the endothelial barrier have evolved over time. To ensure that the normal characteristics of the vascular endothelium are replicated, primary cultured endothelial cells are typically studied, often within a defined number of population doublings after being dissociated from adult or neonatal tissues (See review, Baselet et al. [31]). Immortalized endothelial cells have been used to facilitate the expansion and propagation of cells, but immortalization introduces obligate caveats to the results, as the introduction of mutated or constitutively active signaling proteins can alter gene regulation and signaling in general [83,84]. It has been hypothesized by some researchers that in vitro vascular systems should be derived from stem cells, induced from earliest developmental stages, to differentiate into mature endothelium to best reproduce an endothelial cell system competent for multimodal interrogation [85,86]. Model in vitro endothelial barrier systems have been further improved using co-culture systems with feeder layers or other explanted cell types normally found in the vasculature.

Most recently, the evolution of in vitro endothelial barrier systems has progressed into three-dimensional (3D) systems where tubuloid and spheroid architectures can be assessed by morphometry [87] and the movement of tracer molecules [88]. It is difficult to understate the importance of higher order 3D structure to the system of study. The dependence on cell shape and ordered proximity of various heterologous cell types has been illustrated in the lactogenic function of mammary epithelium [89] and is also seen with human hair production of dermal papilla cells [90]. In vitro reproduction of the tissue architecture is not the only important factor. The vascular milieu in vivo is subjected to shear forces and turbulent flow which differ between vascular beds (artery vs. capillary, for example) that alter endothelial cell anatomy [91–93]. Finally, it is hypothesized that resident endothelial cells in an in vitro model should not be clonal. Endothelial cell heterogeneity is observed in vivo in the expression patterns of junction molecules, with tissues variations [94,95]. Thus, current advances in endothelial barrier studies combine multiple cell types and 3D in vivo cellular architecture with the goal of increasing the fidelity with which in vitro systems recapitulate the “true” endothelial vascular microenvironment.

A variety of experimental techniques have been used to study the elements of the endothelial barrier and its regulating signal transduction. The importance of the components of cell-cell junctions have been revealed by a variety of techniques such co-immunoprecipitation and western blotting, fluorescent chimeric molecules, fluorescent immunohistochemistry, and live cell fluorescent microscopy. The greatest detail of anatomy of the barrier has been revealed in ultrastructural surveys using electron microscopy, especially with the use of immunogold tracers [96–98]. Common assays of the barrier function include monitoring the movement of soluble tracers, such as Alexa Fluor-conjugated ovalbumin or fluorescein-conjugated dextran, through a permeable support surface (transwell system) [99–102]. The greatest resolution in time and degree is afforded when using bioelectric approaches. Transendothelial electrical resistance (TEER) can be measured across monolayers on semipermeable filter inserts or using impedance methods on gold planar arrays using electrical cell impedance sensing technology [99–102].

What, then, are the molecules responsible for barrier function? In vitro studies demonstrated that there are paracellular and transcellular pathways for the movement of ions and biomolecules and transiting leukocytes. The importance of structures involved in lateral cell sealing were first functionally assessed in epithelial cells by scanning an electrode over a cell monolayer and observing increased resistance over cell junctions [103]. A large body of subsequent work has shown that some elements of lateral cell sealing molecules are conserved between epithelial barriers and vascular endothelial barriers. Connections between endothelial cells include tight junctions, adherens junctions, and gap junctions [93,104]. The tight and adherens junctions form the primary lateral sealing structures in endothelial monolayers, while gap junctions allow intercellular chemical and electrical communication [93]. Endothelial tight junctions are populated with junctional adhesion molecules, occludins, and claudins family proteins, linked to the actin cytoskeleton by the Zonula Occludens (ZO) proteins [105–109]. Adherens junctions are comprised predominantly of VE-cadherins, which are linked to the actin cytoskeleton through interactions with α catenin [110–115]. Many other proteins have been identified in endothelial cell-cell junction, including immunoglobin and proline-rich receptor-1 (IGPR-1), intercellular adhesion molecule-1 (ICAM-1), and PECAM [60,116,117]. PECAM-1, a transmembrane Ig superfamily member, engages in homotypic binding with cognate molecules on neighboring endothelial cells and can regulate tight junction integrity [118–121].

In contrast with epithelial cells, endothelial cells overlap each other at cell boundaries, which contributes to the active function of the barrier. The barrier junctions are not arranged in the conventional, stratified belts as seen in polarized epithelial cells, but rather are arranged in radial and intercalated arrays of structures which lend themselves to nonbinary dynamism of the barrier [122].

Several signal transduction pathways have been recognized that maintain barrier homeostasis and that induce permeability. Stabilization of the complex between the cortical cytoskeleton and catenin-VE-cadherin preserves the correct cellular localization of adherens junction proteins and is also required to maintain the half-life of tight junction proteins occludin and claudin [123—126]. Intracellular tension is an important component for junctional stabilization, and signaling to the actin cytoskeleton to cause formation of actin stress fibers and induce actin-myosin contraction can lead to centripetal contraction of endothelial cells, inducing intercellular permeability [127—129]. The rearrangement of the actin cytoskeleton can affect organization of junctional proteins and, by causing a contraction of the cell, weaken the strength of the cell-cell contacts. Vascular permeability can be affected by signaling to modulate a number of actin binding proteins, including cortactin, ezrin/radixin/moesin, vasodilator-stimulated phosphoprotein, etc., that affect actin binding to junctional proteins [93,130]. In addition, the small GTP-binding proteins Rac-1, Cdc42, and Rap-1 were demonstrated to be required for maintenance of adherens junctions and cell polarity; reductions of these proteins resulted in barrier leak [93,127,131]. Signal transduction leading to phosphorylation and activation of myosin by the myosin light chain kinase (MLCK), usually downstream of Ca^{2+}/calmodulin or c-Src activation, can increase endothelial cell contractility and induce permeability [93]. And finally, the small GTP-binding protein RhoA and Rho-associated coiled-coil forming protein kinase (ROCK) function in concert to downregulate myosin light chain phosphatase (MLCP), a phosphatase that can counteract MLCK and activation of myosin [93]. In this way, RhoA/ROCK signaling increases actin-myosin contraction, reducing cell-cell contacts and contributing to vascular permeability. The concerted and/or independent regulation of these signaling pathways can modulate overall vascular permeability.

Vesicular endothelial transcytosis transport, an independent mechanism for endothelial cell-mediated molecular transport, has been most often been studied separately from investigations examining the intercellular endothelial barrier function. Transcytosis occurs via *N*-ethylmaleimide (NEM)-sensitive caveolae [55]. The movement of caveolae requires multiple proteins, including GTP-binding proteins, annexins, the NEM-sensitive fusion protein (NSF), the soluble NSF attachment protein (SNAP), the SNAP receptor (SNARE), and vesicle-associated SNAP receptor (vSNARE), and vesicle-associate membrane proteins (VAMPs) [51,57]. Although these defined pathways have been shown to be involved in intracellular transport, the intersection between transcytosis regulation and paracellular transport regulation has yet to be examined.

How do the endothelial barrier structures and their regulatory apparatus respond to radiation? Studies indicate that radiation-induced reactive oxygen species generated within the cytosol impart inside-out signaling which alters the permeability of the barrier in two ways: (1) through intracellular alteration of the position or composition of the repertoire of junction molecules or their interaction with the cytoskeleton and (2) through extracellular proteolytic remodeling outwardly presented junction molecules (Fig. 2.1).

A common finding from radiation in vitro studies of the endothelial barrier is the reduction in levels of proteins required for endothelial cell-cell contacts. Radiation is known to impart a transient change in the electrical impedance of cultured primary endothelial cell monolayers seeded on collagen within hours after irradiation, followed by waves of change in transmonolayer impedance [65]. This finding has been demonstrated using human umbilical vein endothelial cells (HUVECs) and human brain microvascular endothelial cells [35,102]. The changes correlate with reductions of PECAM-1 protein levels at the lateral junctions. Experiments showed that PECAM-1 protein levels were reduced at endothelial junctions within 3 h following exposure to 5 Gy gamma irradiation, corresponding with a 200 Ω reduction in TEER resistance and increased transit of tracers in transwell inserts [102]. The changes were also observed in 3D in vitro model studies using these cell types [102,132]. Interestingly, particle radiation also reduces endothelial impedance [132,133]—a finding pertinent for spaceflight associated exposures. Reductions of PECAM-1 were also observed in response to 5 Gy photon radiation [132]. Findings from other laboratories further corroborate these findings. 4 Gy irradiation of HUVECs upregulated the cell adhesion protein CD44 that modulates endothelial barrier integrity via regulation of PECAM-1 expression, potentially a regulatory mechanism to compensate for reduced PECAM-1 [134].

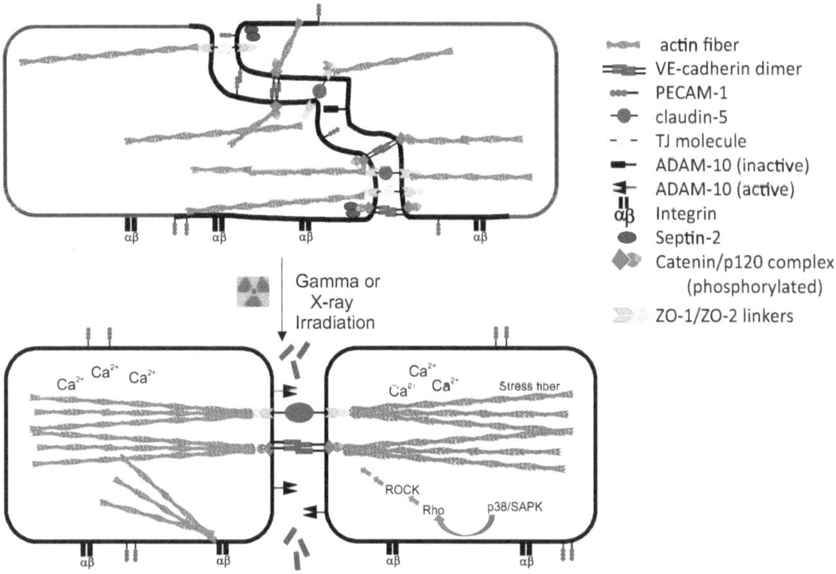

Figure 2.1 ***Mechanisms of radiation-induced changes of endothelial barrier properties.*** The endothelial lateral junctions are comprised of tight junction and adherens junction proteins and septin that stabilize cell-cell overlaps to contribute variously to the integrity of the barrier. Irradiation causes intracellular calcium concentration to increase which results in activation of ADAM10, an extracellular sheddase that proteolytically cleaves VE-cadherin, reducing productive lateral contacts. Reactive oxygen species production results in activation of p38/SAPK signaling culminating in actin rearrangement leading to the formation of stress fibers. This rearrangement causes retraction of PECAM-1 away from the junction and can cause gaps between cells further contributing to reduced permeability of the monolayer.

Radiation-induced modulation of PECAM-1 may be a component of a larger story in paracellular changes that involve the reductions of specific proteins. VE-cadherin and claudin 5 were also demonstrated to be reduced in cultures of human coronary artery endothelial cells (HCAECs) exposed to 0.5–10 Gy X-rays [66]. In a two-dimensional model of cerebrovasculature using HUVECs co-cultured with astrocytes, reduced barrier function was associated with radiation-induced reduction in ZO-1, VE-cadherin, and claudin mRNA and protein levels after exposure to 2–6 Gy (2 Gy/min) [135]. In these studies, cells were seeded on matrigel-coated transwells, and the movement of tracers was monitored. Studies demonstrated that VE-cadherin was cleaved by activation of disintegrin and metalloproteinase family member, ADAM10 following radiation exposure [66]. ADAM10 is known to participate in vasculogenesis and was shown in vitro to facilitate transendothelial migration of leukocytes by modifying VE-cadherin [136,137]. ADAM10 activation by radiation was dependent upon intracellular calcium signaling, but was independent of vascular endothelial growth factor (VEGF) signaling, which is otherwise observed to modulate VE-cadherin stability and has been demonstrated to be altered in vivo by radiation [138,139]. Reduced levels of claudin 5 appear to occur via a different mechanism. Radiation induced dose-dependent decreases in claudin 5 protein expression that were not associated with changes in transcription. In this model, claudin 5 appeared to be constitutively degraded, and radiation exposure either directly or indirectly modulated the degradation. Indeed, claudin 5 protein levels were found to be increased using either a proteosomal or lysosomal inhibitor [66].

Another mechanism for the reduction of endothelial barrier function by radiation is thought to involve intracellular signaling through the RhoA/ROCK signaling pathway [52]. The activation of the Rho/ROCK signaling was linked to a rapid assembly of actin stress fibers and MLCK phosphorylation in irradiated HUVECs, resulting in increased contractility of the cells. In these studies, the increased stress fiber formation was accompanied by VE-cadherin remodeling, in which VE-cadherin was reduced in cell-cell contacts, and tight junctions appeared to be segmented and discontinuous, as shown by immunohistochemistry following radiation. Interestingly, in this model, total VE-cadherin protein was not down-regulated, as determined by western blotting [52].

Radiation-induced permeability changes are not limited to the paracellular route. Changes in cell adhesion and transcellular transport may also contribute. The same group that reported radiation-induced changes in CD44 also identified decreased expression of aquaporin 1 (AQP1; a protein involved in water and hydrogen peroxide balance through transcytosis), the plasminogen activator (PLAT; a protein that helps to dissolve blood clots), and vascular cell adhesion protein 1 (VCAM1; a factor known to participate in leukocyte diapedesis) [134]. Together these gene expression changes are consistent with reduced adhesion and altered molecular transport.

Together, in vitro research has shown that radiation-induced endothelial permeability occurs through at least two mechanisms: (1) the regulation of levels of cell-cell contact proteins and (2) the increased contractility of endothelial cells. Model systems for the study of the endothelial barrier continue to be improved by including more complexity to reflect the in vivo milieu. The molecular anatomy of vascular endothelium is heterogeneiou. Differences are observed in careful comparisons of blood vessels from different tissues. In particular, cellular components differ as do the reperoire of proteins present in cell-cell contacts [140–142]. In addition, most endothelial barrier studies examining the effects of radiation do not take into account fluid flow/shear conditions, and findings from in vitro studies must be considered in context. It is important to note that cells do not represent a mouse, and a mouse very poorly represents a human [143–146]. Sex differences may also play a role in vascular permeability, however, this variable is difficult to reproduce in simplified model systems [147,148]. Finally, the effect of radiation on transcytosis in endothelial cells has been far less studied compared to intercellular barrier function, and the relative contribution of changes to this mechanism of transport to overall permeability changes is not known. Thus, future work with models more representative of the true vasculature may identify additional mechanisms of radiation effects.

2.2 Radiation effects on endothelial cell senescence and survival in vivo and in vitro

Radiation-induced cell death was first recognized in histological samples from early radiation researchers, patients, and volunteers. Studies of animal models showed that cell death in tissues was a function of the dose, dose rate, and total volume of radiation exposure [149]. Radiation activates a variety of signaling pathways in response to the induction of DNA damage, lipid and protein oxidation, endoplasmic reticulum (ER) stress, and redox stress, and the outcome of these signaling events depends strongly upon the cellular context [150,151]. Following radiation exposure, normal cells may undergo a variety of biological responses including accelerated senescence, apoptosis (intrinsic and/or extrinsic), necrosis or necroptosis, or repair and survival [150,151].

2.2.1 Radiation activation of programmed cell death in endothelial cell populations

The effects of radiation on endothelial cell death were first examined in the microvasculature of the skin, where injuries were most readily observed by early radiation researchers. Khan and Ohanian performed electron microscopy studies on the skin capillaries of irradiated rabbit ears, to try to determine the fate of irradiated endothelial cells [36]. At 1–7 days postirradiation, endothelial cell swelling occurred in the microvasculature, with depletion of organelles, dilation of membrane-bound vesicles, and clumping of nuclear chromatin [36]. Interestingly, the images of chromatin "clumping" are consistent with apoptotic nuclear events, although other assays for apoptosis were not available at the time this research was performed.

A comparison of radiation effects on a variety of microvascular beds showed similar responses in the endothelium, with endothelial apoptosis reported in the lung, brain, gastrointestinal (GI) tract, myocardium, and bone marrow [29]. A study of the bone marrow vascular niche, following >5 Gy (0.6 Gy/min) total body irradiation, demonstrated "corruption" of the sinusoidal vascular architecture by 6 and 24 h postirradiation [74]. Bone marrow cells were analyzed by flow cytometry, using Annexin V to identify apoptotic cells, and 7-aminoactinomycin D (7AAD) to identify cells without intact membranes, combined with the panendothelial cell antigen antibody (MECA-32) to identify endothelial cells. Data showed that apoptosis ($Annexin^{+}7AAD^{-}$) occurred at 24 h and 14 days postirradiation, and that necrosis ($Annexin^{+}7AAD^{+}$) occurred at 6 and 24 h postirradiation. Apoptosis and necrosis levels returned to baseline after 14 days postirradiation [74]. Another study by Paris et al. examined radiation effects on the microvasculature of the small intestine [152]. Apoptosis was detected in histological sections using terminal deoxynucleotidyl transferase mediated dUTP nick end labeling (TUNEL) and Annexin V staining, with PECAM-1 counterstaining to identify intestinal endothelial cells [152]. At 4 h following 8–15 Gy irradiation, significant levels of apoptosis were observed [152]. In this study, apoptosis of the lamina propria was demonstrated to precede columnar epithelial apoptosis, and it was hypothesized that endothelial cell apoptosis was a precipitating factor in subsequent damage to other intestinal cells [152]. In support of this, two studies showed that blockade of GI endothelial cell death delayed lethality because of radiogastritis [153,154].

BBB permeability, a well-recognized adverse effect of radiation therapy for cancer treatment in the central nervous system (CNS), was shown to correlate with endothelial cell apoptosis [155]. A study of adult rat subependyma, using TUNEL staining to identify DNA fragmentation characteristic of apoptotic cells, demonstrated the presence of apoptosis in the endothelium within 6–24 h following 0.5–2 Gy radiation exposure [156]. A second study of endothelial cells in the brain using fluorescent immunohistochemistry demonstrated a disappearance of endothelial cells from the rat brain from 1 day through 6 months after exposure to 5–200 Gy radiation [157]. The contribution of endothelial cell apoptosis to BBB permeability was further investigated using transgenic mice, where the inhibition of endothelial cell apoptosis attenuated BBB permeability. In these studies,

BBB permeability was monitored using Evans blue dye leakage around brain microvessels and uptake of the BBB permeability tracer technetium-99m diethylene-triamine-pentaacetic acid (^{99m}Tc-DTPA) at 24 h following 50 Gy irradiation [155]. Importantly, reduced endothelial cell apoptosis was associated with lower levels of neural progenitor cell apoptosis by radiation, suggesting that protection of the endothelium associated with the CNS resulted in improved health of the tissue following radiation exposure [158].

The effect of low dose/high LET charged particle space radiation on the viability of endothelial cells was examined in mice, using immunohistochemistry to examine cellular changes in the ocular microvasculature. After 35 days in space flight, total retinal cells and retinal endothelial cells displayed twice the amount of TUNEL staining as in controls, indicating increased apoptosis [159]. However, this study did not examine other possible contributions of other factors, such as low gravity and changes in ocular pressure. Therefore, additional studies are needed to confirm the effects of low dose/high LET on the vasculature.

The in vivo findings of activation of apoptosis in vascular endothelium led to studies to investigate the mechanism of radiation-induced apoptosis. In vitro studies of cells exposed to DNA damaging agents and redox stress provided evidence that cells could undergo p53-dependent or -independent apoptosis [150,158]. An apoptotic pathway that appears to be commonly utilized in endothelial cells is the p53-independent, ceramide pathway [160]. In this molecular pathway, ceramide is generated from sphingomyelin by two enzymes, acid sphingomyelinase (aSMase) and plasma membrane-bound neutral sphingomyelinase (nSMase). Both enzymes are specialized forms of phospholipase C localized primarily in the plasma membrane that are activated in response to a variety of cell stressors [161]. Once generated from sphingomyelin, ceramide is biologically active, acting as a second messenger molecule in the plasma membrane to form enriched plasma membrane domains, called lipid rafts [161]. The ceramide-containing lipid rafts can capture and cluster death receptors (CD95), creating apoptotic signalosomes that stabilize receptor-ligand interaction complexes and amplify intracellular signaling [160–164]. Internal signaling downstream of activated death receptors can lead to endonuclease activation, Bax activation, and the release of mitochondrial cytochrome *c*, leading to DNA fragmentation [163–165]. Sphingomyelinase has been shown to be activated directly by redox stress [166]. A variety of endothelial cell types (HUVEC, HCAEC, and bovine aortic endothelial cells [BAEC]) were demonstrated to be particularly sensitive to the p53-independent sphingomyelin/ceramide apoptotic pathway in response to redox stress from a variety of stimuli, including ionizing radiation [167–171]. The vulnerability of endothelial cells to ceramide-dependent apoptosis was hypothesized to be because of exceptionally high expression levels of aSMase, up to 20 times the levels of other cell types, which is preferentially trafficked to the plasma membrane and also secreted to the extracellular matrix [172]. The involvement of redox stress-induced SMase-dependent endothelial cell death was confirmed in in vivo studies using aSMase deficient mice [173].

Radiation, through the generation of reactive oxygen species (ROS), was also demonstrated to activate the ceramide-dependent pathway of apoptosis. Hamovitz-Freidman et al. provided evidence that single dose irradiation (500 cGy) of cultured BAEC induced the activation of nSMase, resulting in rapid sphingomyelin hydrolysis to ceramide that preceded DNA laddering and apoptosis at 24 h (Fig. 2.2) [174]. A ~30% reduction in sphingomyelin and ~30% increase in ceramide could be detected within 1 min of radiation exposure [174–176]. Earlier studies of radiation-induced apoptosis in BAEC demonstrated that treatment with bFGF inhibited cell death in a protein-kinase C (PKC)-dependent manner. Interestingly, basic fibroblast growth factor (bFGF) did not affect radiation-induced DNA damage or the rate of DNA repair, suggesting that the change in survival was because of an effect on apoptotic signaling. Based on the inhibition of endothelial cell apoptosis by PCK activation, Hamovitz-Friedman et al. investigated the role of PKC in the ceramide-dependent apoptotic pathway. Activation of PKC by phorbol esters inhibited sphingomyelin degradation to ceramide by radiation [174]. Interestingly, experiments also demonstrated that radiation could increase the activation of nSMase in nuclei-free cellular extracts, suggesting that nuclear components and signals originating from DNA damage were not required for initial sphingomyelinase activity [174].

Liao et al. demonstrated that radiation inhibited ceramide synthase (CS) (Fig. 2.2) [177]. In these studies, BAEC were exposed to ^{125}I-labeled deoxyuridine or X-irradiation (2.5–15 Gy) to induce apoptosis, and ceramide, SMase and CS were all measured. CS, which controls de novo synthesis of ceramide, was chemically inhibited using the *Fusarium moniliforme* toxin fumonisin B1 (FB1), a natural inhibitor of CS [177]. The ceramide production from breakdown of sphingomyelin, via activation of plasma membrane SMase, was not affected by FB1. However, FB1 blocked the de novo synthesis of ceramide by inhibiting CS and subsequent

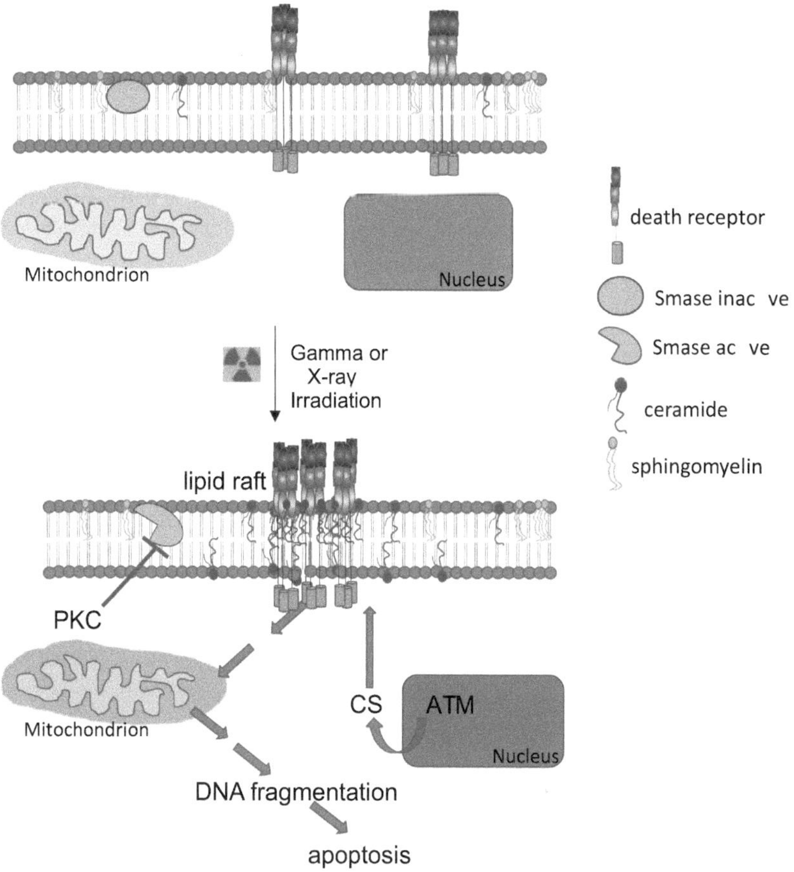

Figure 2.2 ***Mechanism of radiation-induced apoptosis in endothelial cells.*** Radiation exposure can result in the rapid activation of SMase, an enzyme that proteolytically cleaves sphingomyelin to generate ceramide. Increased plasma membrane levels of ceramide can generate lipid rafts where the clustering of receptors can occur, including death receptors, resulting in more efficient activation. In addition, DNA breaks induced by radiation leads to activation of the DNA damage response, including the activation of ataxia telangiectasia-mutated that can upregulate ceramide synthase. Death receptor activation signals to induce the extrinsic pathway of apoptosis. Activation of protein-kinase C can inhibit downstream signaling for apoptosis.

radiation-induced DNA damage-induced apoptosis. In these studies, CS was shown to be activated by radiation at in a dose-dependent manner (peaking at 10 Gy), beginning at 3–4 h, with maximal activation at 12 h [177]. It was concluded that ceramide could be generated de novo in response to radiation via CS activation. Interestingly, this activation was not inhibited by cyclohexamide, suggesting that CS itself was activated by radiation and was not synthesized from new protein translation. These studies also investigated the effect of radiation on immortalized B cell lines from patients with ataxia telangiectasia-mutated (ATM). These patients are known to exhibit compromised responses to radiation damage, as ATM transduces some of the most rapid signaling events following the induction of DNA breaks. B cells normally display ceramide-dependent apoptosis in response to radiation, but B cells deficient in ATM were found to be resistant to radiation-induced apoptosis. These experiments showed that in the presence of mutated ATM, radiation-induced CS activation was blocked [177]. In contrast, inhibition of caspases did not block CS activation. CS activation after radiation required intact DNA damage response, in contrast with SMase that are activated directly by radiation [177].

Human cells with mutations in the ceramide pathway enzymes and knockout mice deficient in enzymes for the production of ceramide were used to confirm the requirement of ceramide for radiation-induced apoptosis in vivo. Niemann-Pick patients, who have an inherited deficiency of aSMase activity, have defects in fat metabolism and fail to produce ceramide from sphingomyelin. Studies showed that lymphoblasts from Niemann-Pick patients failed to undergo apoptosis in response to ionizing radiation, also consistent with a link between ceramide production and apoptosis [167]. In aSMase knockout mice, radiation exposure failed to induce the proapoptotic lipid ceramide, and peripheral microvascular endothelial cells were resistant to radiation-induced apoptosis in the intestine and brain endothelium [155,167,178]. The restoration of aSMase resulted in increased radiation sensitivity [179]. Together these findings provided strong evidence for the specificity of the sphingomyelin/ceramide pathway for endothelial apoptosis in response to radiation in vivo. Interestingly, p53 knockout mice were also used to study radiation-induced apoptosis in the GI and bone marrow. Data showed that, in contrast with radiation-resistant aSMase knockout mice, the p53 knockout mice had no change in levels of radiation-induced endothelial cell apoptosis in the GI [152]. In addition, $p53^{-/-}$ mice displayed *increased* endothelial cell loss, decreased microvascular density, and accelerated mortality following 12 Gy irradiation of the heart [180].

In contrast with studies of cultured aortic or systemic (nonpulmonary) endothelial cells, studies using cultured human or bovine pulmonary artery endothelial cells (PAECs) demonstrated that significant levels of apoptosis could not be detected below 30–50 Gy X-irradiation [181]. Following exposure to 50 Gy X-ray irradiation, PAEC can undergo intrinsic or extrinsic apoptosis, as indicated by caspase 8 and 9 activation. Exposure to 50 Gy ionizing radiation activated the ER stress response and upregulation of the transcription factor X-box binding protein 1 and its downstream target glucose regulated protein 78,000 (GRP78), an ER chaperone. Suppression of the ER stress response with salubrinal reduced X-ray-induced apoptosis in PAECs. The dependence of apoptosis upon SMase activation, ceramide, or p53 activation was not examined. It is not known whether the ER stress pathway is a functional pathway downstream of ceramide production or if this is an independent mechanism of apoptosis activated in endothelial cells from the pulmonary artery. The findings suggest that there is a difference between the response of nonpulmonary and pulmonary derived endothelial cells to ionizing radiation. Interestingly, although the endothelium throughout the body has many similarities, regional differences exist in morphology and physiology, including hemodynamics and signaling responses [182]. The pulmonary artery vasculature is derived from different embryonic tissues from the systemic endothelium, display differences in signal transduction, and responds differently to pharmaceutical agents [183]. The difference between radiation responses for the endothelial cell from the two vascular systems provides another example for differences between these two systems.

Together, the studies of ionizing radiation-induced apoptosis suggest that in systemic (nonpulmonary) microvascular endothelial cells and arterial endothelial cells, ceramide-dependent, p53-independent apoptosis is a predominant response, both in vitro and in vivo [184]. Further studies are needed to determine the underlying mechanism(s) for differences in sensitivity to radiation-induced apoptosis and whether ceramide signaling is required in all endothelial cell types. In addition, differences in the radiation responses of the pulmonary and nonpulmonary vascular systems suggest that different agents may be required for the prevention of radiation damage.

2.2.2 Mechanisms of radiation-induced accelerated senescence in endothelial cells

Cellular senescence, or replicative senescence, describes the phenomenon of irreversible cell cycle and growth arrest that occurs in most cells [185,186]. Normal human diploid cells can usually replicate ~40–60 times (termed the Hayflick Limit) before undergoing natural senescence that is believed to be controlled by telomere shortening over time [185]. However, under conditions of cellular stress, such as redox stress, DNA damage, or in the presence of some oncogenic mutations, cells can enter into accelerated senescence prior to reaching their normal replicative lifespan [185,187]. Senescent cells remain metabolically active, but have altered gene and protein expression, have an abnormal secretome (the senescence-associated secretory phenotype, SASP), exhibit changes in cell-cell interactions, and are resistant to apoptosis. The presence of senescent cells is currently regarded as a problem in aging or damaged tissues, as the accumulation of senescent cells can alter tissue structure and function, promoting inflammation and oncogenesis [186,188].

Two major signaling pathways have been demonstrated to lead to replicative and stress-induced accelerated senescence: the p53-p21/Waf1 pathway (sometimes called telomere-dependent) and the telomere-independent p38 mitogen activated protein kinase ($p38^{MAPK}$) pathway [186,187]. Both pathways lead to the increased expression of cell cycle checkpoint proteins and converge on the retinoblastoma (Rb) protein that controls G_1 to S phase progression. Senescence-associated beta galactosidase (SA-β-gal) is used as a marker in both cases [187,189]. Replicative senescence (telomere length-dependent) and DNA damage-induced senescence are believed to proceed primarily through the activation of ataxia telangiectasia mutated (ATM) protein leading to increased levels of tumor suppressor protein p53 and upregulation of the cyclin-dependent kinase inhibitor p21/Waf1 protein. In contrast, the telomere-independent pathway of senescence requires activation of $p38^{MAPK}$ that can upregulate cytoplasmic p16/Ink4 cyclin-dependent kinase inhibitor/tumor suppressor, leading to Rb inhibition of cell cycle progression [187]. Oxidative stress can activate both pathways, by the induction of DNA damage response and telomere erosion or by the direct activation of $p38^{MAPK}$ [187]. Both pathways also lead to the activation of nuclear factor-kappa B (NF-κB), which is considered the primary transcription factor for the regulation of SASP [186].

In vitro studies showed that acute exposure to high-dose gamma radiation induced dose-dependent senescence in adult endothelial cells. In most cases, senescence was confirmed by assaying for SA-β-gal, often in combination with western blotting for upregulation of cell cycle checkpoint proteins. In HUVECs, senescence occurred within 5 days postirradiation in a dose-dependent manner: 2 Gy led to ~30% senescence, 4 Gy led to ~50% senescence, and 8 Gy led to ~80% senescence [190]. In a separate study of radiation-induced gene expression changes, 4 Gy irradiation of HUVECs resulted in changes in expression of several genes associated with cell cycle regulation, usually p21/Waf1 and p16/Ink4a, consistent with the induction of senescence [134]. In rat cerebromicrovascular endothelial cells (CMVECs), radiation-induced senescence was shown to be dose dependent, with ~10% senescence at 2 Gy irradiation and ~30% senescence at 4–8 Gy, measured at 7 days postirradiation [191]. Interestingly, in CMVECs, senescence still occurred although DNA was active and double-stranded DNA breaks were repaired to near baseline levels, suggesting that although DNA damage was resolved, senescence was irreversibly induced [191]. In contrast with HUVECs and CMVECs, cultured bovine PAEC appear to be more sensitive to the induction of senescence, 75% senescence was observed with as low as 5 Gy X-ray irradiation [192].

The induction of senescence in vitro was also shown to be dependent upon the dose rate, and not just total dose of radiation exposure. HUVECs exposed to chronic low-dose radiation at 2.4–4.1 mGy/h for 10 weeks displayed accelerated senescence as indicated by expression of p21/Waf1 and SA-β-gal [193,194]. Reverse phase protein arrays and triplex Isotope Coded Protein Labeling with LC-ESI-MS/MS was used to show that the PI3K/Akt/mTOR pathway was inactivated during senescence, and that pathways related to DNA repair, oxidative stress, and inflammation were influenced [193,194]. However, when exposed to chronic low dose 1.4 mGy/h for 10 weeks, HUVECs showed increased p21/Waf1 without SA-β-gal, indicating the activation of cell cycle arrest without advancement to senescence [193]. This suggests that a threshold level for either redox stress or DNA damage must be reached to induce final signaling for senescence. (Note, the existence of a threshold for the induction of senescence does not correlate with a lack of DNA damage or potential health risks for low-dose radiation exposure [195,196].)

Cell culture studies of radiation effects on human lung microvascular endothelial cells (HLMVEC) provided evidence for the requirement of p53 activation for senescence (Fig. 2.3). In HLMVEC, senescence induced by radiation (1–15 Gy) could be blocked by pifithrin-α, a p53 inhibitor [197]. An elegant set of experiments in HLMVEC examined the role of the X-linked inhibitor of apoptosis (XIAP)-associated factor 1 (XAF1) for regulating p53 and senescence by radiation. XIAP is a ubiquitin E3 ligase that can suppress cell death by inducing the breakdown of caspases 3, 7, and 9 as well as other cell death regulators [198]. XAF1 is a tumor suppressor that antagonizes XIAP. Since XIAP suppresses cell death, the upregulation of XAF1 leads to increased cell death, by binding XIAP and blocking the suppression of caspase activity [199]. XAF1 has also been demonstrated to regulate senescence [200]. Studies showed that radiation resulted in upregulated XAF1 in HLMVEC. Furthermore, knockdown of XAF1 using

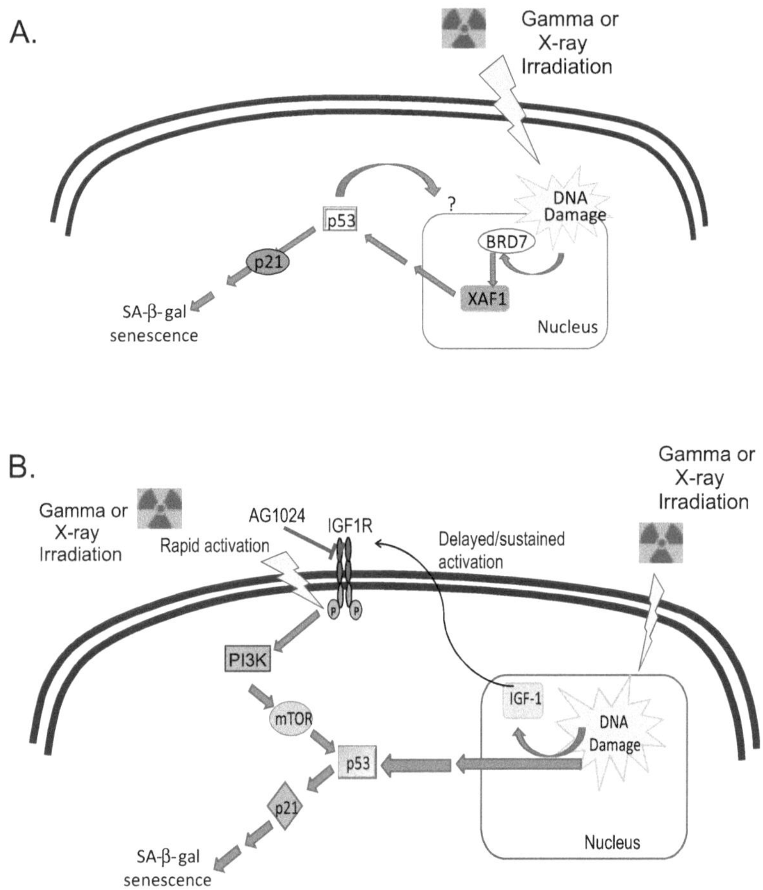

Figure 2.3 ***Mechanisms of radiation-induced senescence in endothelial cells.*** (A) Radiation-induced DNA damage can result in the activation of the BRD7 transcription factor that can regulate expression of XAF1. The activation of XAF1 results in activation of p53 to induce downstream signaling resulting in senescence. (B) IGF1R activation by ionizing radiation is necessary for radiation-induced accelerated senescence in some endothelial cell types. Ionizing radiation-induced DNA damage activates p53 and p21/Waf1 (p21). The cell cycle is blocked by the activation of p21/Waf1 and other cell cycle checkpoint proteins. IGF-1R receptor is rapidly activated by ionizing radiation, and can signal through PI3K and mTOR, also affecting p53 activation and leading to senescence. AG1024, an inhibitor of IGF-1R, suppresses radiation-induced senescence.

siRNA suppressed the activation of p53 by radiation, suggesting that XAF1 was required in signaling for p53 activation and senescence. Overexpression of XAF1 alone in HLMVEC was shown to be sufficient to induce senescence. XAF1-induced senescence was examined in the presence of shRNA retroviruses targeting p53 and p16Ink4A [201]. These experiments demonstrated that knockdown of p53 reduced the level of senescence induced by XAF1 overexpression, whereas knockdown of p16Ink4A did not. Final experiments examined the regulation of XAF1 by the DNA-damage response transcription factor bromodomain containing 7 (BRD7), and provided evidence for BRD7 activation for XAF1 expression following radiation exposure [201]. Together these data provide a potential pathway for p53-p21/Waf1 pathway activation that requires the DNA damage response and regulation of XAF-1.

Senescence in bovine and human PAEC exposed to 10 Gy X-ray irradiation was demonstrated to require the activation of the insulin-like growth factor receptor (IGF1R) for the activation of p53 [192] (Fig. 2.3). In PAEC, radiation led to the immediate activation of IGF1R and subsequently induced endogenous expression of insulin-like growth factor-1 (IGF-1), an IGF1R ligand, thus inducing an autocrine cycle. IGF1R signaling induced increased levels of p53, upregulation of p21/Waf1, and subsequent expression of SA-β-gal [192]. Inhibition of IGF1R, or inhibition of downstream signaling molecules mammalian target of rapamycin (mTOR) or PI3K led to the suppression of p53 upregulation, p21/Waf1 upregulation, and SA-β-gal expression following radiation exposure. In addition, cells maintained a normal morphology in the presence of the IGF1R inhibitor [192]. The identification of the IGF-1 pathway for radiation-induced senescence correlated to senescence pathways in aging-associated senescence, where IGF-1 signaling was also identified [187]. The interactions between the IGF1R pathway of senescence and the XAF1/BRD7 pathway are not yet identified.

As stated above, a primary characteristic of senescent cells is altered synthesis and secretion of a variety of proteins, including extracellular matrix proteins, proteases, cytokines, and growth factors [187,202]. The overall effect is the establishment of the senescence-associated secretory phenotype, which is frequently proinflammatory [187,202]. The regulation of senescence is believed to be upstream of activation of the senescence-associated secretory phenotype. Several studies have demonstrated that the NF-κB transcription factor regulates the expression of a number of proinflammatory genes [203]. Both DNA damage and oxidative stress can activate NF-κB, leading to the SASP [203]. The role of NF-κB in radiation SASP was examined in HUVECs, where 8 Gy irradiation-induced NF-κB levels and activity, as well as IL-6 expression, an indicator of SASP [190]. DNA damage was shown to activate NF-κB through the regulation of NF-κB essential modulator (NEMO) [204]. The NEMO inhibitor PS1145 transiently reduced by radiation-induced expression of IL-6, a cytokine characteristically expressed in the SASP [190]. However, PS1145 reduced the levels of SA-β-gal and increased the percentage of cells undergoing apoptosis [190]. The reduction of senescence was associated with suppression of radiation-induced p53-induced death domain (PIDD), a protein with multiple proteolytic fragments implicated in cell cycle arrest, senescence, and apoptosis [205]. These data suggest that NF-kB may participate in a feed-forward loop for the induction of specific PIDD isoforms, senescence, prevention of apoptosis, and establishment of the senescence-associated secretory phenotype [190].

Radiation-induced cell cycle arrest, which can occur independently from senescence, was demonstrated in vivo in several tissues. An early study by Imaizumi et al. investigated the effects of 20 Gy radiation on implanted aortic rings in mice [206]. Researchers observed the upregulation of p53 and p21/Waf1 following radiation exposure in the endothelial layer of the implanted tissue. The endothelium exhibited a growth arrested state that was no longer responsive to VEGF- or fibroblast growth factor-2 (FGF-2)-induced proliferation [206]. Interestingly, the researchers also found that radiation-induced senescence could be prevented by inhibition of the transforming growth factor beta (TGF-β) type 1 receptor ALK5, suggesting a role for TGF-β signaling in growth arrest [206]. While this study did not examine an endogenous endothelial bed, it established a senescence-like response of endothelial cells in the environment following ionizing radiation in vivo. Studies by Himburg et al. provided evidence that radiation-induced (5–7 Gy) changes in cell cycle regulating gene expression occurred in endogenous endothelial cells in murine bone marrow and brain tissues [74]. In the bone marrow, within 6–24 h of radiation exposure, alterations were observed in the expression of cell cycle arrest genes cyclin-dependent kinase 18 (Cdk18) and cell division associated 7 (Cdca7) [74]. In the brain, an elevation was observed in cyclin-dependent kinase inhibitor 1A (Cdkn1a) at 6 h–7 d post-irradiation [207]. Nominally, radiation exposure results in cell cycle arrest and increased apoptosis, but this was not found to be the case in the brain, where cell cycle arrest was not accompanied by increased apoptosis; caspase 3 activation was not found and apoptosis-associated gene expression programs were not induced for the acid or nSMase pathways [207,208]. These studies suggested that the cell cycle was arrested by radiation but was not accompanied by apoptosis. However, senescence was not specifically examined.

Evidence for senescence was also obtained from in vivo studies of tumor vasculature and normal brain vasculature. In a study of brain endothelium in glioma tumors, 6 Gy radiation induced endothelial expression of gamma-H2A.X (γ-H2AX), a variant of histone H2A that is increased following double-stranded DNA breaks and is associated with senescence and the expression of SA-β-gal [187,209]. Studies of irradiation of normal brain tissue provided evidence of endothelial senescence with the expression of SA-β-gal together with the upregulation of cell cycle regulating proteins p16INK4A and p21/Waf1 in the cerebrum and cerebellum with 2–6 days of 2–20 Gy exposure [187,210,211]. Low and moderate dose cranial irradiation (0.1 and 2 Gy, respectively) also resulted in significant endothelial senescence in the brains of mice (10-day-old and 10-week-old), as determined by SA-β-gal staining [210]. Radiation-induced senescence was found to be associated with the reduction of synthesis of junction proteins claudins, ZO-1, and β-catenin, suggesting a link between senescence and changes in permeability [210]. Although ionizing gamma and X-ray irradiation were demonstrated to induce apoptosis in the BBB endothelium, senescence was also identified [210]. In the brain, vascular leak and endothelial senescence by radiation was associated with reduced immunostaining of tight junction protein claudin-5 and glial fibrillary acidic protein (GFAP) in the cell-cell junctions [210].

In vivo, bone marrow vascular endothelial cells were shown to contain activated p53 at 24 h postirradiation after 1–3 Gy total body irradiation. In this study, p53 activation was hypothesized to lead to cellular senescence, as apoptosis could not be detected [212]. Senescence was also strongly implicated in studies of late (16-week) vascular injuries following 8–16 Gy cardiac irradiation in mice [213]. These studies utilized proteomics, transcriptomics, and bioinformatics to show that radiation exposure impaired energy metabolism and altered insulin/insulin growth factor/phosphatidyl-inositol 3 kinase-Akt signaling [213]. Together these findings are indicative of radiation-induced endothelial senescence in the vasculature of multiple tissues. In most cases, increased endothelial senescence was associated with reduced vascular barrier function.

Together, in vivo and in vitro studies of radiation effects on cell survival provided evidence of the p53-p21/Waf1 dependent pathway of senescence in endothelial cells. Additional mechanisms for the regulation of p53 (IGF1R or NEMO) require further investigation to determine whether these pathways are interrelated, and whether they are activated in all types of endothelial cells. As stated above, the differences in radiation responses endothelial cells from different vascular beds could be because of underlying differences in the endothelial cells themselves [182].

2.3 Radiation induction of activated pro-inflammatory endothelial cells

Endothelial cells can be stimulated to assume an "activated" phenotype. In their activated state, endothelial cells participate in the generation of inflammation by increasing surface expression of cell adhesion proteins and by the secretion of pro-inflammatory chemokines and cytokines for immune cell activation and attraction [28,50]. Endothelial cells can be activated by a variety of insults, including stress (physical or redox stress), exposure to pathogens (bacteria or viruses), or exposure to other agents such as specific cytokines or growth factors, or autoimmune antiendothelial antibodies [49,214]. The induction of senescence in endothelial cells, described above, also results in an altered cell type with increased production of pro-inflammatory factors, the senescence-associated secretory phenotype [30,187,189,215]. In vivo, senescence of the endothelium can result in a proinflammatory phenotype and has been associated with age-related cardiovascular diseases [149]. The secretions of activated endothelial cells can increase local inflammation and exacerbate loss of barrier function and dysregulation of vascular tone [28].

In vivo studies showed that endogenous endothelial cells became activated after radiation exposure of >2 Gy, characterized by increased production of inflammatory cell chemoattractants and adhesion molecules [31]. Following 10 Gy cranial irradiation in mice, the cerebral endothelial cells were separated by fluorescence activated cell sorting and used for gene array analysis. Consistent with in vitro findings, the in vivo radiation responses from 6 h through seven days postirradiation endothelial cells exhibit increased expression of genes encoding ICAM-1 and VCAM-1 [207,211]. CMVECs irradiated (2–8 Gy) in culture underwent accelerated senescence and initiated secretion of pro-inflammatory cytokines: interleukin (IL)-1α, IL-6, tumor necrosis factor-alpha (TNF-α), C–C motif chemokine-11 (Ccl11; eotaxin-1), and C–C motif chemokine ligand 2 (Ccl2; monocyte chemoattractant protein 1 [MCP-1]) [30,191,207]. Microvascular endothelial cells in the skin respond in a similar fashion. At a higher dose of 10 Gy, radiation caused increased expression of adhesion proteins E-selectin, ICAM-1, VCAM-1, and CD44 [30,216]. The combination of upregulation of cellular adhesion molecules and immune chemoattractants in the vascular endothelium is hypothesized to contribute to subsequent leukocyte sequestration in irradiated tissues [217].

The activation of redox stress- and DNA damage-responsive transcription factors, nuclear factor (erythroid-derived 2)-like 2 (Nrf2), activator protein

1 (AP-1), and nuclear factor-kappaB (NF-κB), are thought to be a primary mechanism for radiation-induced endothelial responses [28,31,50,218]. Redox stress leads to Nrf2 translocation to the nucleus where it binds to the antioxidant response element (ARE) to increase gene expression a variety of antioxidant enzymes [219]. AP 1 activation in endothelial cells increases proliferative and proinflammation gene expression, although its role in radiation-induced endothelial cell activation has not been completely elucidated [219]. Activation of NF-κB is believed to regulate most of the cytokines and adhesion molecules expressed in endothelial cells [28,220]. NF-κB has been shown to increase expression of adhesion molecules E-selectin and ICAM in irradiated endothelium. The proinflammatory pathway activated by radiation in endothelial cells was demonstrated to be dependent upon the regulation of NEMO, cytoplasmic activator of NF-κB [190]. Irradiation of HUVECs in the presence of a small molecule inhibitor of NEMO suppressed NF-κB regulation of IL-6 [190].

An alternative mechanism for the development of the proinflammatory phenotype of irradiated endothelial cells is the suppression of anti-inflammatory proteins. Using a murine brain-derived endothelial cell line, bEnd.3, McRobb et al. demonstrated that radiation-induced accelerated senescence resulted in the suppression of disintegrin and metalloproteinase domain-containing protein 10 (ADAM10), a protease responsible for the cleavage of multiple proteins involved in microvascular and neural homeostasis, including the amyloid precursor protein (APP) to its soluble neuroprotective alpha fragment sAAPα [211]. Whereas sAAPα suppresses inflammation following cleavage, the full length APP is proinflammatory with toxic effects on the endothelium [221]. Researchers observed the accumulation of other proinflammatory ADAM10 target proteins, including neural cell adhesion molecule L1 (L1CAM), nestin (NEST), neogenin (NEO1), Tolllike receptor 2 (TLR2), and an ATP-dependent RNA helicase, retinoic acid inducible gene 1 (RIG-1), several of which are key activators of neuro-inflammation and innate immunity. Together, the increased levels of these proinflammatory proteins could have further effects on endothelial barrier permeability in the brain and the systemic micro-vasculature [211,214].

Together, findings indicate that development of activated endothelial cells following radiation exposure involves increased expression of proinflammatory cytokines, increased cell surface expression of adhesion molecules for the recruitment of leukocytes, and decreased anti-inflammatory pathways. The combination of increased adhesion proteins and chemoattractants is significant, as cytokines serve to attract and activate white blood cells while the increased expression of the adhesion proteins contributes to their attachment to the endothelium in a first step toward diapedesis. Leukocyte accumulation has been observed in animal models of radiation injury in the lung, liver, and kidney, and is believed to be a significant late pathological indicator [217,222–224]. Blockade of endothelial cell activation and inhibition of leukocyte transmigration could be a viable target for pharmaceutical intervention for mitigation or treatment of late radiation injuries [217].

3. Indirect mechanisms of radiation-induced vascular effects

Direct effects of radiation on the endothelial cell of the vascular barrier can lead to specific signal transduction, inducing alterations in cell-cell contacts as well as endothelial cell survival, as noted above. However, radiation can also impact the function of endothelial cells through indirect mechanisms, resulting in delayed effects on vascular leak and edema. These effects include: (1) the induction of a proinflammatory state in the vasculature environment through the activation of mast cells, or, in the case of the BBB, the activation of astrocytes and (2) radiation-induced hemolysis of reticulocytes and the release of iron. Each of these subsequent events can have independent and long-term effects on barrier function and vascular tone.

3.1 Radiation-induced inflammation and vascular permeability

Inflammation has a broad effect on vascular permeability and vascular tone to allow wide spread signaling for immune cell activation and the passage of immune cells to sites of infection or tissue damage through diapedesis [49]. Both local and systemic inflammatory processes modulate the vasculature to achieve effective immune cell infiltration of tissues.

The immune system is sensitive to radiation, and exposure to even low levels of radiation can activate immune cells, causing the release of pr-inflammatory cytokines from hematopoietic cells and nonhematopoietic tissues [37,225,226]. Early inflammatory responses to radiation can be generated by white blood cells, including mast cells, which degranulate in response to radiation exposure. Soon afterward, these immune cells die. High-dose total body radiation exposure induces a loss of mature white blood cells and many hematopoietic progenitors from apoptosis, leading to immune suppression [227]. Interestingly, a delayed inflammatory response can be generated following acute high-dose radiation exposure, including the robust production of proinflammatory cytokines, despite the deficiency of mature white blood cells [228–232]. The source of delayed cytokine production is not completely understood, but radiation-induced proinflammatory responses are believed to participate in subsequent tissue damage even outside of the immediate radiation field [232–234]. Long-term studies of atomic bomb survivors and radiotherapy patients identified increased levels of cytokines in the blood, suggesting a long-term alteration of the overall inflammatory state [30,37].

3.1.1 Mast cell activation in radiation-induced vascular permeability

Mast cells are present in all vascularized tissues and are abundant in the skin [235]. Mast cells are prolific in their production of proinflammatory factors, which include VEGF, leukotrienes (LTs) B_4, C_4, D_4, and E_4, prostaglandin D_2 (PGD_2), cathepsin G, IL-1, IL-8, tryptase (α and β), platelet activating factor (PAF), TNF-α, bFGF, and transforming growth factor-beta (TGF-β) [235]. The released cytokines and proteases can alter local blood flow, increase vascular permeability, and initiate cascades of proinflammatory events and tissue remodeling [67,225,236–241]. Substances released by mast cells increase local inflammation as well as systemic inflammation by the delivery of inflammatory factors into the blood circulation [235].

In the early 1940's–1950's, researchers examined histological samples to study inflammatory mast cell activation and degranulation, and similarities were identified for radiation effects on these cells. Sylven observed alterations of the mast cell structures, loss of granules and cellular swelling, suggestive of degranulation in histological sections of carcinoma tissue from patients treated with X-rays [242]. Skin sections from irradiated rats provided evidence that radiation induced biphasic mast cell degranulation, with a short, early phase at 0.5–1 h postirradiation followed by a prolonged second phase lasting up to 10 days, with a peak from 5 to 12 h postirradiation [243,244]. Histamine, known to be an abundant substance within mast cell granules, was quantified in tissues to confirm mast cell activation [63]. Studies in rats, rabbits, hamsters, and chicks provided evidence that X-ray irradiation caused a decrease in tissue levels of histamine in the skin, stomach, jejunum, and lung [63,245]. At the same time, histamine was decreased in the tissues; it was observed that histamine levels in the serum were increased. In irradiated human patients, the systemic release of histamine was accompanied by reduction in blood pressure, a major physiological effect of histamine [246]. Together these findings suggested that there was a mechanistic link between radiation-induced mast cell degranulation, increase systemic histamine, and changes in vasoconstriction.

Later studies used pharmacological inhibitors to identify substances responsible for radiation-induced vascular permeability. Antihistamines were used to treat inflammation as early as the late 1940's, and based on evidence of mast cell degranulation by radiation, the use of antihistamines was initially advocated for the prevention of skin damage during radiotherapy [247,248]. Mixed results were obtained using antihistamines to reduce radiation-induced vascular leak. By monitoring the release of trypan blue or I^{125}-labeled albumin, it was shown that antihistamines reduced local vascular permeability induced by radiation in rat intestines and rabbit ears [249–251]. However, other studies showed greater inhibition of vascular leak using protease inhibitors [70]. Jolles and Harrison showed that treatment of the skin with several inhibitors (soybean trypsin inhibitor, ovomucoid trypsin inhibitor, and ε-amino-n-caproic acid [an inhibitor of carboxypeptidase B]) could block radiation-induced vascular permeability [70]. Several other agents described as "compounds with vascular action" (hexopal and plaquenil) were also shown to suppress radiation-induced vascular effects [70]. At this time, the complete components of mast cell granules had not been identified, and the targets of protease inhibitors were not known.

In later studies used transgenic mice deficient in mast cells to investigate the link between mast cell activation and capillary permeability. Comparisons between mast cell deficient WBB6F1/J-Kit^{W}/Kit^{W-v} mice and their wild type littermates mice used intravenously injected fluorescent dye (fluorescent isothiocyanate [FITC]-labeled dextran) to image the microvasculature by intravital microscopy for radiation-induced vascular permeability [241]. This technique allowed the calculation of total volume and surface area of microvessels [241]. In addition, injected fluorescently labeled bovine serum albumin allowed the calculation of permeability-induced protein loss from the serum, using the rate of change of intensity in imaging with the known rate of albumin clearance under normal vascular conditions [241]. Experiments showed that vascular permeability at 24 h postirradiation was suppressed in mast cell deficient mice [241]. This study showed that there were no changes in the expression of NO or VEGF during the time course of vascular permeabilization. Although these studies provided strong evidence for the role of mast cell activation in radiation-induced alteration of vascular permeability, the agents released by mast cells to affect the endothelial barrier were not identified [241].

Mast cell population numbers were also affected by radiation. Mast cells populations were demonstrated to initially increase in tissues within 24 h of radiation, followed by a decrease by ~5 days post-irradiation [63], although the source of the cellular increase was not determined. Mast cells at these time points were described as swollen with less tightly packed granules, and some with ruptured membranes [63], The nadir in mast cell numbers was shown to occur ~10 days postirradiation, with cell numbers recovery after 16 days postirradiation. Differences between exact times of mast cell increases and decreases in different studies may reflect differences in the dose of radiation exposure and the type of tissue examined. Late effect studies in mice and of humans exposed to nuclear events showed that following radiation exposure mast cell numbers were eventually increased [63,252].

Together, these findings suggest that radiation effects on mast cells influence both vascular leak and long-term vascular remodeling. Degranulation and activation of mast cells results in both immediate and delayed effects on vascular permeability. In addition, mast cell populations are biphasically regulated following radiation, with initial decreases followed by steady increases. Findings suggest that mast cell populations in the long term do not return to normal homeostatic levels, potentially resulting in a persistent proinflammatory state that may contribute to long-term changes in inflammation and tissue dysfunction following radiation exposure.

3.1.2 Astrocyte activation by ionizing radiation

The specialized vascular barrier that separates the CNS stringently regulates all molecular exchanges between the cerebrovascular interstitial fluid and the blood [253]. The specialized endothelial cells that comprise the BBB are linked by tight junctions that occlude the intercellular cleft between adjacent endothelial cells and together with different types of specialized cells form a sophisticated neuro-function system that has also been defined as the "neurovascular unit" [254]. The tight junctions of the BBB restrict even small ion movements, and most nutrients for the brain are therefore imported through receptor-mediated or adsorptive-mediated transcytosis through the endothelial cells. Indeed, the TEER of the brain endothelium is $>1000\ \Omega\ cm^2$, a very resistive barrier compared with peripheral capillaries which have a TEER of some $\sim 2-20\ \Omega\ cm^2$ [253].

Astrocytes, a glial cell type comprising 30% of the CNS, participates in the regulation of BBB permeability [253,255]. Of the eleven identified astrocyte phenotypes, eight have been shown to interact with blood vessels of the CNS in strategic positions between the endothelium and neurons [253,256]. The microcapillary endothelial cells of the BBB are surrounded by perivascular astrocyte "endfeet," that participate in the production of the extracellular matrix of the local basal lamina as well as modulate the flow of ions, amino acids, neurotransmitters, and water [253]. The astrocyte endfeet processes secrete signaling molecules to regulate endothelial cells that affect vasodilation and tight junction protein expression [257]. Astrocytes communicate with each other through several mechanisms, including gap junctions, which are believed to function to coordinate their regulation of the endothelial barrier [258]. The perivascular astrocytes also display specific protein expression (including the water channel aquaporin 4 (AQP4), the K^+ channel Kir4.1, the Ca^{2+}-dependent $rSloK_{Ca}$ channels, and proton transporters) that contribute to electrolyte and water homeostasis in the extracellular space [253,259]. Thus, astrocytes regulate BBB permeability through multiple mechanisms, through the regulation of gene expression in the endothelial cells as well as through expression of specific transport proteins on their cellular membranes.

Inflammation in the brain and spinal column alters the interactions between astrocytes and endothelial cells of the BBB [253,260,261]. The CNS was originally thought to be excluded from immune cells of the rest of the body, a phenomenon called "immune-privileged," based on early studies demonstrating the lack of immune surveillance function for tumor detection in the CNS. More recent studies demonstrated that lymphatic drainage exists for the CNS, and immune cells such as T cells can be detected in some regions of the brain [262]. However, a major part of inflammatory responses in the CNS involve resident glial cells, including microglia, considered to be the myeloid cells of the brain, and astrocytes [255]. Astrocyte activation can be induced by a variety of insults including trauma, microbial and viral infection, and neurodegenerative diseases [255,263]. Cytokines that can induce astrocyte activation include TNF-α and IL-1α [255]. The activated astrocytes have altered morphology, displaying hypertrophy and outgrowth of long processes, and can display increased proliferation [264,265]. Activated astrocytes also have increased expression of inflammatory mediators including IL-1β and TNF-α [255]. The GFAP is a common marker for normal astrocytes, but has increased expression with astrocyte activation [266].

BBB permeability is a proxy for endothelial cell adhesion, lateral sealing, transcellular transport, and apoptosis [210]. Dead and missing cells clearly do not support barrier function. Senescent cells similarly exhibit an incomplete contribution to barrier function. However, studies show that in addition to direct damage to the endothelium, radiation causes the activation of astrocytes. Although astrocytes themselves are resistant to radiation- and redox-induced senescence and apoptosis [267,268], radiation can induce astrogliosis. Intriguingly, some studies show that astrogliosis could be introduced by microglia activation [269]. In vivo, exposure of whole rat brain to 15 Gy irradiation resulted in astrogliosis within 6 h, as revealed by increased GFAP [269]. Similar changes were observed in astrocytes following fractionated irradiation (2 Gy daily irradiation 5 days/week, 40 Gy total) of whole brain in mice [62]. In this study, increased vascular permeability (indicated by fluorescein isothylcyanate-dextran permeability and intravital microscopy) was detected ~90 days postirradiation, with astrogliosis detectable at ~60 days postirradiation [62]. In vivo irradiation of mouse brains (delivered as 7 fractions of 5 Gy, 35 Gy total) showed that a wide range of alterations occurred simultaneously, including BBB permeability, microglial activation, astrocyte activation, and expression of inflammatory proteins IL-1β, TNF-α, and COX-2 [270]. In vitro studies showed that co-culture of endothelial cells derived from the BBB with activated astrocytes resulted in decreased barrier function as determined by TEER and increased expression of cytokines and chemokines [271]. The decreased in barrier function induced by the activated astrocytes was associated with reduced expression of ZO-1 and claudin proteins on the endothelial cells [261].

The mechanisms of astrocyte activation subsequent to irradiation have been investigated in vitro and in vivo. In vitro, irradiation of astrocytes alone did not result in astrocyte activation, but a mixed culture of astrocyte-microglia cell preparation demonstrated astrocyte phenotype changes consistent with astrogliosis [269]. Experiments were performed to demonstrate that irradiation of microglia induced the expression of prostaglandin E2, which could signal to astrocytes to induce astrogliosis-associated morphological changes [269]. In a separate in vivo study, the role of the radiation-induced cytokines were investigated for their role in vascular membrane leak and astrogliosis [272]. Treatment of mice with anti-TNF-α antibodies, administered 15 min before and 23 h after irradiation, reduced BBB vascular permeability, microvessel dilation, and leukocyte adhesion at 24 and 48 h following 20 Gy localized brain irradiation. In these studies, anti-TNF-α antibodies also reduced the level of astrogliosis at 24 and 48 h postirradiation [272].

The exact mechanism by which activated astrocytes change endothelial barrier function is not completely understood. The existing findings suggest that suppression of astrogliosis following brain irradiation improves vascular barrier function and decreases brain damage. Radiation-induced astrocyte activation may involve several cell types, including microglia, suggesting a complexity in the mechanism. Further insight into the mechanism(s) of astrocyte-modulated vascular permeability could provide additional drug targets for preventing radiation neuronal injury following clinical radiation treatment.

3.2 Radiation-induced hemolysis and the effect of iron overload on the endothelial barrier in vivo

Iron is a vital nutrient with critical roles at the cellular level in enzymatic reactions, mitochondrial respiration and is essential in oxygen transport for mammals [273]. The processes of initial acquisition of iron from the diet and its distribution are regulated by specific cytokines in response to blood iron concentrations, with detection within the liver to regulate absorption in the small intestine [274,275]. The most active usage of iron in mammals occurs in the bone marrow for erythropoiesis, and 65%—75% of total iron is found in heme associated with hemoglobin within erythrocytes [275]. Macrophages are a central cell type in this process, obtaining transferrin-(TF-) bound iron from the plasma to process for storage of iron within the bone marrow to provide to developing erythroblasts. Macrophages also recycle iron by ingesting senescent or damaged red blood cells (RBCs), degrading their hemoglobin, and binding the Fe to ferritin in the cytoplasm, usually within the spleen [275,276].

Iron absorption and transport are tightly regulated because unbound iron can promote the generation of toxic free radicals through its interaction with oxygen, as described in the Fenton and Haber-Weiss reactions [274,275,277,278]. Dysregulated iron accumulation and iron-catalyzed free radical generation (Fig. 2.1) have been proposed to play a role in cancer, cardiovascular disease, metabolic disease, blood disorders, and multiple degenerative diseases [278—281]. Excessive levels of iron also negatively impact the hematopoietic system. Dietary iron overload in a murine model was demonstrated to impair the ratio and clonogenic function of hematopoietic stem and progenitor cells, concomitant with increased levels of reactive oxygen species generation [282]. Excessive hemolysis of RBC, by cellular damage or under other pathological conditions, can trigger increased expression of iron-binding and transport proteins in macrophages [275]. Surplus iron uptake alters gene expression in macrophages such that normal antimicrobial activities are inhibited, and can induce a pathological state in macrophages, such as occurs in atherosclerosis [276,281,283].

The hematopoietic system is sensitive to damage by ionizing radiation, and exposure to high-dose total body irradiation [284,285]. In response to ionizing radiation, white blood cells and many hematopoietic progenitors primarily undergo apoptosis in response to DNA damage [227,286]. However, reticulocytes and RBC are resistant to radiation-induced apoptosis because of their lack of DNA and the absence of apoptotic machinery [287]. In reticulocytes and RBC, radiation induces hemolysis as a result of hemoglobin oxidation and denaturation as well as morphological changes associated with altered mechanical properties of the cytoskeleton and membrane [286–288]. Radiation-induced hemolysis and release of iron may play a role in radiation injury independent from the loss of erythroid cells themselves [275,282,289]. In murine models of total body irradiation, serum iron was shown to be significantly elevated within one day following irradiation, and to continue to increase through 20 days postirradiation [290,291]. In addition, direct irradiation of the femur alone in mice was also shown to increase serum and liver iron levels [292]. Importantly, the chelation of iron following irradiation of one femur in mice was demonstrated to positively impact bone damage by suppressing excessive osteoclast activity and restoring balance with osteoblast homeostasis [292]. However, additional studies are required to define the mechanism(s) of the toxic effects of iron on hematopoiesis.

Several studies have indicated that iron has direct effects on endothelial cell function. In cultured bovine aortic endothelials cells, the application of iron as iron sucrose (0.1–1 mg/mL) decreased proliferation and increased p53-dependent apoptosis as the result of DNA damage [293]. Interestingly, iron dextran at the same concentrations had less toxic effects on cultured endothelial cells, with reduced induction of DNA damage and apoptosis; the reason for this difference was unexplored, but it suggests that not all iron complexes are equally toxic to cells [293]. Iron ($FeCl_3$, 300 μM) in HUVEC cultures induced the generation of hydrogen peroxide and hydroxyl radicals, detected by 2′,7′-dichlorodihydrofluorescein diacetate, and induced mitochondrial damage and apoptosis [294]. Moreover, the effects of $FeCl_3$ were mitigated by co-incubation with the iron chelator deferiprone, which blocked iron-induced ROS generation and decreased mitochondrial damage and apoptosis [294].

The presence of extracellular iron was also demonstrated to be sufficient to reduce endothelial barrier function in culture [295]. In cultured brain vascular endothelial cells, incubation with free iron ($FeSO_4$, 30–300 μM), iron-bound hemoglobin (which contains Fe^{2+}, 1–25 μM), or hemin (an Fe^{3+}-containing porphyrin with chlorine, 1–50 μM) induced ROS production and apoptosis associated with intracellular accumulation of Fe^{2+} [295]. Evaluation of the endothelial barrier by trans-endothelial resistance and FITC-dextran permeability showed that incubation with hemin also increased barrier permeability, which was associated with decreased VE-cadherin staining [295].

In the blood, iron-containing heme is present at ~20 mM, but it is contained within erythrocytes and does not directly contact other cell types [296]. Free iron can be liberated from the heme complex in an enzyme-dependent reaction performed by heme oxygenase-1 (HO-1) [296]. Fogg et al. found that endogenous HO-1 is basally expressed in rat and human PAECs. HO-1 is upregulated by the presence of heme and free iron ($FeSO_4$, 1 μM), and, to a lesser extent, by hyperoxia [297]. Furthermore, treatment of PAECs with deferrioxamine, an iron chelator, inhibited iron-dependent, heme-dependent, and hyperoxia-dependent induction of HO-1 expression, and reduced HO-1 to below basal expression levels [297]. It is thought that the induction of HO-1 expression could be a cytoprotective response in endothelial cells in the presence of iron, as HO-1 can reduce ER stress in the presence of heme and promote the clearing of heme by releasing unbound iron for binding to ferritin and recycling [298].

The idea that the vasculature in vivo is susceptible to cytotoxic effects of free iron from lysed erythrocytes seems intuitive based on the data from iron effects on cultured endothelial cells. In vivo studies of the vascular effects of iron have examined endothelial cell function in maintenance of the BBB. Brain endothelial cells contain relatively high levels of ferritin and act as iron reservoirs for brain tissue [299]. Disruption of this BBB occurs in various forms of brain injury, which can result in iron deposition in the brain (as either heme or nonheme), that can result in redox stress [300]. In transient forebrain ischemia induced in mice, Won et al. showed that iron-mediated endothelial damage increased BBB permeability [301]. Region-specific iron overload in the brain was linked to the production of free radicals following forebrain ischemia [301]. Importantly, the injection of iron-bound hemin alone was sufficient to damage the BBB and cause vascular leak as well as reduced expression of occludin and vascular endothelial-cadherin [295]. Data from studies of transient forebrain ischemia provided evidence that iron-mediated free radicals in the brain were responsible for neuronal death and gliosis following transient forebrain ischemia [302,303]. The iron chelators deferoxamine and 2,2′-bipyridil blocked free radical production, reduced BBB permeability and endothelial cell degeneration following transient forebrain ischemia or hemin injection [295,301].

Together, these studies suggest that iron has a significant impact on vascular barrier permeability.

The release of iron by ionizing radiation exposure in vivo, from the rapid hemolysis of erythrocytes, can cause a cascade of adverse effects on endothelial cell survival and barrier function. These effects of iron are independent of the immediate effects of radiation exposure and may result in compounding initial radiation damage to the endothelium.

4. Conclusions

The understanding of radiation-induced health effects has advanced significantly over the ~120 years since the initial discovery of this energy source. However, the potential for radiation exposure has also dramatically increased in this century. Radiotherapy is currently a first line treatment for many cancers, and is a co-therapy modality for most cancers [304]. With the proliferation of nuclear power plants and nuclear weapons, there is also an increasing risk of accidental radiation exposure [305,306]. In addition, preparation for human participation in space exploration, with a focus on extended space missions, has highlighted the need for protection against heavy ion radiation [133]. Research suggests that radiation effects on the vasculature may underlie early, delayed, and late tissue injuries [21,29,31,33–37]. Furthermore, the network of cell–cell junctions that are unique among endothelial cells of the vasculature may create an avenue whereby radiation damage and cellular responses can easily be transferred through the cellular layer, potentially resulting in bystander effects, through which signals may be propagated [307–309]. Fortunately, research continues to improve the understanding of the mechanisms of radiation damage to the vascular endothelium, and findings from these studies will likely contribute to the development of pharmaceutical and biological radiation countermeasures for tissue protection from clinical or accidental exposures.

Acknowledgments

This work was supported by a pilot grant from the Opportunity Funds Management Core of the Centers for Medical Countermeasures against Radiation, National Institutes of Health (NIH), National Institutes of Allergy and Infectious Diseases, grant number U19AI067773 (Award Number 5U19AI067773-14, P.I. Regina M. Day), by Award Number DM178018 (P.I. Regina M. Day) from the Defense Medical Research and Materiel Command, Radiation Health Effects Research Program, Joint Program Committee 7, and by the NIH, National Cancer Institute grant R01 CA184168 (P.I. Albert J Fornace Jr, former

P.I. Elliott Rosen; Co-investigator Regina M. Day). Some of the authors are employees of the US Government, and this work was prepared as part of their official duties. Title 17 U S C. §105 provides that "Copyright protection under this title is not available for any work of the United States Government." Title 17 U S C §101 defines a US Government work as a work prepared by a military service member or employees of the US Government as part of that person's official duties. The views in this article are those of the authors and do not necessarily reflect the views, official policy, or position of the National Institute of Allergy and Infectious Diseases or the National Institutes of Health, the Uniformed Services University of the Health Sciences, the Armed Forces Radiobiology Research Institute, Department of the Navy, Department of Defense or the US Federal Government.

References

[1] Posner E. Reception of Rontgen's discovery in Britain and U.S.A. Br Med J 1970; 4(5731):357–60.
[2] Rontgen W. Eine neue art von Strahlen. Nature 1895;53:274.
[3] Becquerel A. On the invisible rays emitted by phosphorescent bodies. Comptes Rendus 1896;122:501–3.
[4] Sansare K, Khanna V, Karjodkar F. Early victims of X-rays: a tribute and current perception. Dentomaxillofacial Radiol 2011;40(2):123–5.
[5] Stevens LG. Injurious effects on the skin. In: report on the application of the new photography to medicine and surgery. Br Med J 1986;1:998.
[6] Munk J, Peyser E, Gruszkiewicz J. Radiation induced intracranial meningiomas. Clin Radiol 1969;20:90–4.
[7] Moss W, Eckhardt R. The human plutonium injection experiments. Los Alamos Sci 1995;23:177–233.
[8] Lisner P. Beitrag zur histologie der roentgenwirkung auf die normale meschliche haut. Fortschr a d Geb d Roentgenstrahlen 1905;8:97–100.
[9] Gassman A. Zur histologic roentgen-ulcera. Fortschor Roentgenstr 1899;2:199–211.
[10] Unna P. Die chronische rontgendermatitis der radiologen. Forschr Gebiete Rontgenstrahlen 1904;8:67–91.
[11] Fajardo LF, Berthrong M, Anderson RE. Radiation Pathology. New York: Oxford University Press; 2001.
[12] Simpson FE. Radium therapy. St. Louis: C.V. Mosby Co; 1922.
[13] Pullinger BD. Causes of death in irradiated human tissue. J Pathol Bacteriol 1932;35: 527–40.
[14] Historical summary. Acta Radiol 1955;43:13–23.
[15] Vugman I. Release of histamine by noxious physical agents, organic compounds and by metals. In: Eichler O, Farah A, editors. Histamine and Anti-Histaminics - Part 1: Histamine. Its Chemistry, Metabolism and Physiological and Pharmacological Actions. Berlin: Springer-Verlag; 1966. p. 367–85.
[16] Jolles B. The reciprocal vicinity effect of irradiated tissues on a "diffusible substance" in irradiated tissues. Br J Radiol 1950;23:18–24.
[17] Smith DE, Svihla G, Patt HM. The effects of X-radiation on circulation in the wing of the bat. Physiol Zool 1951;24:249–57.
[18] Baermann G, Linser P. Uber die lokale und allgemeine wirkund der roentgenstrahlen. Munchen Med Wochenshr 1904;51:966–99.
[19] Rost JA. Biologische wirkung der rontgenstrahlen von verschiedener qualitat auf die haut von mensch und tier. Strahlentherapie Bd 1915;6:269.
[20] Wolbach SB. Pathologic histology of chronic x-ray dermatitis and early x-ray carcinoma. J Media Res 1909;21:415–49.

[21] Satyamitra MM, DiCarlo AL, Taliaferro L. Understanding the pathophysiology and challenges of development of medical countermeasures for radiation-induced vascular/endothelial cell injuries: report of a NIAID workshop, August 20, 2015. Radiat Res 2016;186(2):99–111.
[22] Fajardo LF, Stewart JR. Pathogenesis of radiation-induced myocardial fibrosis. Lab Invest 1973;29:244–57.
[23] Maynard KI, et al. X-irradiation attenuates relaxant responses in the rabbit ear artery. Br J Pharmacol 1992;105(1):126–8.
[24] Hatoum OA, et al. Radiation induces endothelial dysfunction in murine intestinal arterioles via enhanced production of reactive oxygen species. Arterioscler Thromb Vasc Biol 2006;26(2):287–94.
[25] Shimizu Y, et al. Radiation exposure and circulatory disease risk: Hiroshima and Nagasaki atomic bomb survivor data, 1950-2003. BMJ 2010;340:b5349.
[26] Moseeva MB, et al. Risks of circulatory diseases among Mayak PA workers with radiation doses estimated using the improved Mayak worker dosimetry system 2008. Radiat Environ Biophys 2014;53:469–77.
[27] Darby SC, et al. Risk of ischemic heart disease in women after radiotherapy for breast cancer. N Engl J Med 2013;368(11):987–98.
[28] Baselet B, et al. Selected endothelial responses after ionizing radiation exposure. In: Lenasi H, editor. Endothelial Dysfunction: Old Concepts and New Challenges. London: IntechOpen Limited; 2018.
[29] Korpela E, Liu SK. Endothelial perturbations and therapeutic strategies in normal tissue radiation damage. Radiat Oncol 2014;9:266.
[30] Venkatesulu BP, et al. Radiation-induced endothelial vascular injury: a review of possible mechanisms. JACC Basic Transl Sci 2018;3(4):563–72.
[31] Baselet B, et al. Pathological effects of ionizing radiation: endothelial activation and dysfunction. Cell Mol Life Sci 2019;76(4):699–728.
[32] Borak J. Radiation effects on blood vessels. Part I: erythema; edema. Radiology 1942; 38:481–92.
[33] Weintraub NL, Jones WK, Manka D. Understanding radiation-induced vascular disease. J Am Coll Cardiol 2010;55:1.
[34] Phillips SA, Hatoum OA, Gutterman DD. The mechanism of flow-induced dilation in human adipose arterioles involves hydrogen peroxide during CAD. Am J Physiol Heart Circ Physiol 2007;292(1):H93–100.
[35] Young EF. Transient impedance changes in venous endothelial monolayers as a biological radiation dosimetry response. J Electr Bioimp 2012;3:61–5.
[36] Khan MY, Ohanian M. Radiation-induced cardiomyopathy. II. An electron microscopic study of myocardial microvasculature. Am J Pathol 1974;74:125–36.
[37] Halle M, et al. Sustained inflammation due to nuclear factor-kappa B activation in irradiated human arteries. J Am Coll Cardiol 2010;55(12):1227–36.
[38] Citrin D, et al. Radioprotectors and mitigators of radiation-induced normal tissue injury. Oncologist 2010;15(4):360–71.
[39] Hall EJ, Giaccia AJ. Radiobiology for the radiologist. 8th ed. Philadelphia: Wolters Klewer; 2019.
[40] Folkes LK, O'Neill P. Modification of DNA damage mechanisms by nitric oxide during ionizing radiation. Free Radic Biol Med 2013;58:14–25.
[41] Izquierdo-Alvarez A, et al. Identification of S-nitrosylated and reversibly oxidized proteins by fluorescence switch and complementary techniques. Methods Mol Biol 2018;1747:73–87.
[42] Barshishat-Kupper M, et al. Protein oxidation in the lungs of C57BL/6J mice following X-irradiation. Proteomes 2015;3(3):249–65.

[43] Barshishat-Kupper M, et al. Effect of ionizing radiation on liver protein oxidation and metabolic function in C57BL/6J mice. Int J Radiat Biol 2014;90(12):1169–78.
[44] Kim YC, et al. Bone marrow protein oxidation in response to ionizing radiation in C57BL/6J Mice. Proteomes 2014;2(3):291–302.
[45] Reisz JA, et al. Effects of ionizing radiation on biological molecules–mechanisms of damage and emerging methods of detection. Antioxidants Redox Signal 2014; 21(2):260–92.
[46] Houee-Levin C, Bobrowski K. The use of the methods of radiolysis to explore the mechanisms of free radical modifications in proteins. J Proteom 2013;92:51–62.
[47] Risau W. Development and differentiation of endothelium. Kidney Int Suppl 1998; 67:S3–6.
[48] Schnittler H. Between sealing and leakiness: molecular dynamics of the endothelium to maintain and regulate barrier function. Cell Tissue Res 2014;355:481–3.
[49] Sandoo A, et al. The endothelium and its role in regulating vascular tone. Open Cardiovasc Med J 2010;4:302–12.
[50] Hunt BJ, Jurd KM. Endothelial cell activation. A central pathophysiological process. BMJ 1998;316(7141):1328–9.
[51] Stan RV. Channels across endothelial cells. In: Madame curie bioscience database [internet]. Austin (TX): Landes Bioscience; 2000–2013.
[52] Gabrys D, et al. Radiation effects on the cytoskeleton of endothelial cells and endothelial monolayer permeability. Int J Radiat Oncol Biol Phys 2007;69(5):1553–62.
[53] Shakhov AS, Dugina VB, Alieva IB. Structural features of actin cytoskeleton required for endotheliocyte barrier function. Biochemistry 2019;84(4):358–69.
[54] Radeva MY, Waschke J. Mind the gap: mechanisms regulating the endothelial barrier. Acta Physiol 2018;222(1).
[55] Minshall RD, et al. Vesicle formation and trafficking in endothelial cells and regulation of endothelial barrier function. Histochem Cell Biol 2002;117(2):105–12.
[56] Fung KYY, Fairn GD, Lee WL. Transcellular vesicular transport in epithelial and endothelial cells: challenges and opportunities. Traffic 2018;19(1):5–18.
[57] Schnitzer JE, Liu J, Oh P. Endothelial caveolae have the molecular transport machinery for vesicle budding, docking, and fusion including VAMP, NSF, SNAP, annexins, and GTPases. J Biol Chem 1995;270(24):14399–404.
[58] Zhang X, Sessa WC, Fernandez-Hernando C. Endothelial transcytosis of lipoproteins in atherosclerosis. Front Cardiovasc Med 2018;5:130.
[59] Pulgar VM. Transcytosis to cross the blood brain barrier, new advancements and challenges. Front Neurosci 2018;12:1019.
[60] Rahimi N. Defenders and challengers of endothelial barrier function. Front Immunol 2017;8:1847.
[61] Bari WA, Sorenson GD. Ultrastructural alterations in x-irradiated spleen. Pathol Microbiol 1964;27:257–75.
[62] Yuan H, et al. Effects of fractionated radiation on the brain vasculature in a murine model: blood-brain barrier permeability, astrocyte proliferation, and ultrastructural changes. Int J Radiat Oncol Biol Phys 2006;66(3):860–6.
[63] Eisen VD, Ellis RE, Wilson CWM. The effect of X-irradiation on tissue histamine in the rat. J Physiol 1956;133:506–19.
[64] Baker DG, Krochak RJ. The response of the microvascular system to radiation: a review. Cancer Invest 1989;7(3):287–94.
[65] Young EF, Smilenov LB. Impedance-based surveillance of transient permeability changes in coronary endothelial monolayers after exposure to ionizing radiation. Radiat Res 2011;176(4):415–24.
[66] Kabacik S, Raj K. Ionising radiation increases permeability of endothelium through ADAM10-mediated cleavage of VE-cadherin. Oncotarget 2017;8(47):82049–63.

[67] Miles AA, Miles EM. Vascular reactions to histamine, histamin-liberator and leukotaxine in the skin of guinea-pigs. J Physiol 1952;118:228–57.
[68] Jolles B, Harrison RG. Radiation skin reaction and depletion and restoration of body immune response. Nature 1963;198:1216–7.
[69] Siegal T, et al. Cellular and secretory mechanisms related to delayed radiation-induced microvessel dysfunction in the spinal cord of rats. Int J Radiat Oncol Biol Phys 1996;36(3):649–59.
[70] Jolles B, Harrison RG. Enzymic processes and vascular changes in the skin radiation reaction. Br J Radiol 1966;39:12–8.
[71] Jolles B, Harrison RG. Proteases and the depletion and restoration of skin responsiveness to radiation. Nature 1965;205:920–1.
[72] Marchesi VT, Barrnett RJ. The demonstration of enzymatic activity in pinocytic vesicles of blood capillaries with electron microscopy. J Cell Biol 1963;17:547–56.
[73] Teneff S, Stoppani F. L'influenza delle radiazioni sulle linfoghiandole e sulla circolazione linfatica. Radiol Med 1938;22:768–75.
[74] Himburg HA, et al. A molecular profile of the endothelial cell response to ionizing radiation. Radiat Res 2016;186(2):141–52.
[75] Griem ML, Robotewskyj A, Nagel RH. Potential vascular damage from radiation in the space environment. Adv Space Res 1994;14(10):555–63.
[76] Looney WB, et al. A clinical investigation of the chronic effects of radium salts administered therapeutically (1915-1923). Am J Roentgenol Radium Ther Nucl Med 1955;73:1006–37.
[77] Hasterlik RJ. The delayed toxicity of radium deposited in the skeleton of human beings. Proc Int Conf Peaceful Uses Atomic Energy 1955;11:149–55.
[78] Lo EH, Fabrikant JI. Delayed biologic reactions to sterotactic charged-particle radiosurgery in the human brain. Sterotact Funct Neurosurg 1991;56:197–212.
[79] Cox AB, et al. Effects of heavy ions on rabbit tissues: damage to the forebrain. Int J Radiat Biol 1982;42:355–67.
[80] Levy RP, et al. Clinical results of stereotactic heavy-charged particle radiosugery for intracranial angiographically occult vascular malformations. In: Steiner L, editor. Radiosugery: baseline and trends. New York: Raven Press; 1992. p. 149–54.
[81] Levy RP, et al. Computed tomography slice-by-slice target-volume delineation for stereotactic proton irradiation of large intracranial arteriovenous malformations: an iterative approach using angiography, computed tomography, and magnetic resonance imaging. Int J Radiat Oncol Biol Phys 1996;35:555–64.
[82] Lo EH, et al. MRI and PET of delayed heavy-ion radiation injury to the rabbit brain. Int J Radiat Oncol Biol Phys 1991;20:689–96.
[83] Doerfler W, Weber S, Naumann A. Inheritable epigenetic response towards foreign DNA entry by mammalian host cells: a guardian of genomic stability. Epigenetics 2018;13(12):1141–53.
[84] Kamranvar SA, Masucci MG. Regulation of telomere homeostasis during Epstein-Barr virus infection and immortalization. Viruses 2017;9(8).
[85] Park TE, et al. Hypoxia-enhanced blood-brain barrier chip recapitulates human barrier function and shuttling of drugs and antibodies. Nat Commun 2019;10(1):2621.
[86] Weidenfeller C, Svendsen CN, Shusta EV. Differentiating embryonic neural progenitor cells induce blood-brain barrier properties. J Neurochem 2007;101(2):555–65.
[87] Davis GE, Camarillo CW. An alpha 2 beta 1 integrin-dependent pinocytic mechanism involving intracellular vacuole formation and coalescence regulates capillary lumen and tube formation in three-dimensional collagen matrix. Exp Cell Res 1996;224(1):39–51.
[88] Cho CF, et al. Blood-brain-barrier spheroids as an in vitro screening platform for brain-penetrating agents. Nat Commun 2017;8:15623.

[89] Streuli CH, Bailey N, Bissell MJ. Control of mammary epithelial differentiation: basement membrane induces tissue-specific gene expression in the absence of cell-cell interaction and morphological polarity. J Cell Biol 1991;115(5):1383–95.
[90] Abaci HE, et al. Tissue engineering of human hair follicles using a biomimetic developmental approach. Nat Commun 2018;9(1):5301.
[91] Franke RP, et al. Induction of human vascular endothelial stress fibres by fluid shear stress. Nature 1984;307(5952):648–9.
[92] Vereycken-Holler V, Aigueperse J, Gaugler MH. Radiation effects on circulating and endothelial cell interactions studied by quantitative real-time videomicroscopy. Int J Radiat Biol 2002;78(10):923–30.
[93] Komarova YA, et al. Protein interactions at endothelial junctions and signaling mechanisms regulating endothelial permeability. Circ Res 2017;120(1):179–206.
[94] Hanske S, et al. Different segments of the cerebral vasculature reveal specific endothelial specifications, while tight junction proteins appear equally distributed. Brain Struct Funct 2017;222(3):1179–92.
[95] Paul D, et al. Novel 3D analysis of claudin-5 reveals significant endothelial heterogeneity among CNS microvessels. Microvasc Res 2013;86:1–10.
[96] Clough G. Relationship between microvascular permeability and ultrastructure. Prog Biophys Mol Biol 1991;55(1):47–69.
[97] Curry FE, Clough GF. Flow-dependent changes in microvascular permeability – an important adaptive phenomenon. J Physiol 2002;543(Pt 3):729.
[98] Vorbrodt AW, Dobrogowska DH. Molecular anatomy of intercellular junctions in brain endothelial and epithelial barriers: electron microscopist's view. Brain Res Brain Res Rev 2003;42(3):221–42.
[99] Stolwijk JA, et al. Impedance analysis of GPCR-mediated changes in endothelial barrier function: overview and fundamental considerations for stable and reproducible measurements. Pflügers Archiv 2015;467(10):2193–218.
[100] Wegener J, Seebach J. Experimental tools to monitor the dynamics of endothelial barrier function: a survey of in vitro approaches. Cell Tissue Res 2014;355(3):485–514.
[101] Srinivasan B, et al. TEER measurement techniques for in vitro barrier model systems. J Lab Autom 2015;20(2):107–26.
[102] Sharma P, Templin T, Grabham P. Short term effects of gamma radiation on endothelial barrier function: uncoupling of PECAM-1. Microvasc Res 2013;86:11–20.
[103] Gitter AH, Fromm M, Schulzke JD. Impedance analysis for the determination of epithelial and subepithelial resistance in intestinal tissues. J Biochem Biophys Methods 1998;37(1–2):35–46.
[104] Vestweber D. Molecular mechanisms that control endothelial cell contacts. J Pathol 2000;190(3):281–91.
[105] Anderson JM, Balda MS, Fanning AS. The structure and regulation of tight junctions. Curr Opin Cell Biol 1993;5(5):772–8.
[106] Tanaka H, et al. Site-specific distribution of claudin-based paracellular channels with roles in biological fluid flow and metabolism. Ann N Y Acad Sci 2017;1405(1):44–52.
[107] Tsukita S, Furuse M. The structure and function of claudins, cell adhesion molecules at tight junctions. Ann N Y Acad Sci 2000;915:129–35.
[108] Tsukita S, Furuse M, Itoh M. Multifunctional strands in tight junctions. Nat Rev Mol Cell Biol 2001;2(4):285–93.
[109] Wolburg H, Lippoldt A. Tight junctions of the blood-brain barrier: development, composition and regulation. Vasc Pharmacol 2002;38(6):323–37.
[110] Bazzoni G, Dejana E. Endothelial cell-to-cell junctions: molecular organization and role in vascular homeostasis. Physiol Rev 2004;84(3):869–901.

[111] Dejana E. Endothelial adherens junctions: implications in the control of vascular permeability and angiogenesis. J Clin Invest 1996;98(9):1949–53.
[112] Dejana E. Endothelial adherens junctions: implications in the control of vascular permeability and angiogenesis. J Clin Invest 1997;100(11 Suppl. 1):S7–10.
[113] Gumbiner BM. Regulation of cadherin adhesive activity. J Cell Biol 2000;148(3): 399–404.
[114] Yap AS, Brieher WM, Gumbiner BM. Molecular and functional analysis of cadherin-based adherens junctions. Annu Rev Cell Dev Biol 1997;13:119–46.
[115] Kim J, Cooper JA. Septins regulate junctional integrity of endothelial monolayers. Mol Biol Cell 2018;29(13):1693–703.
[116] Heemskerk N, et al. F-actin-rich contractile endothelial pores prevent vascular leakage during leukocyte diapedesis through local RhoA signalling. Nat Commun 2016;7:10493.
[117] van Buul JD, Timmerman I. Small Rho GTPase-mediated actin dynamics at endothelial adherens junctions. Small GTPases 2016;7(1):21–31.
[118] Muller WA. Transendothelial migration: unifying principles from the endothelial perspective. Immunol Rev 2016;273(1):61–75.
[119] Vestweber D. Relevance of endothelial junctions in leukocyte extravasation and vascular permeability. Ann N Y Acad Sci 2012;1257:184–92.
[120] Vestweber D. How leukocytes cross the vascular endothelium. Nat Rev Immunol 2015;15(11):692–704.
[121] Privratsky JR, Newman PJ. PECAM-1: regulator of endothelial junctional integrity. Cell Tissue Res 2014;355(3):607–19.
[122] Carvey PM, Hendey B, Monahan AJ. The blood-brain barrier in neurodegenerative disease: a rhetorical perspective. J Neurochem 2009;111(2):291–314.
[123] Fukuhara S, et al. Cyclic AMP potentiates vascular endothelial cadherin-mediated cell-cell contact to enhance endothelial barrier function through an Epac-Rap1 signaling pathway. Mol Cell Biol 2005;25(1):136–46.
[124] Lampugnani MG, et al. Vascular endothelial cadherin controls VEGFR-2 internalization and signaling from intracellular compartments. J Cell Biol 2006;174(4): 593–604.
[125] Ramirez SH, et al. Inhibition of glycogen synthase kinase 3beta promotes tight junction stability in brain endothelial cells by half-life extension of occludin and claudin-5. PLoS One 2013;8(2):e55972.
[126] Yamamoto H, et al. Integrin beta1 controls VE-cadherin localization and blood vessel stability. Nat Commun 2015;6:6429.
[127] Zhang XE, Adderley SP, Breslin JW. Activation of RhoA, but not Rac1, mediates early stages of S1P-induced endothelial barrier enhancement. PLoS One 2016; 11(5):e0155490.
[128] Lum H, Malik AB. Mechanisms of increased endothelial permeability. Can J Physiol Pharmacol 1996;74(7):787–800.
[129] Manaenko A, et al. Inhibition of stress fiber formation preserves blood-brain barrier after intracerebral hemorrhage in mice. J Cerebr Blood Flow Metabol 2018;38(1): 87–102.
[130] Garcia-Ponce A, et al. The role of actin-binding proteins in the control of endothelial barrier integrity. Thromb Haemostasis 2015;113(1):20–36.
[131] Spindler V, et al. Ultrastructural analysis reveals cAMP-dependent enhancement of microvascular endothelial barrier functions via Rac1-mediated reorganization of intercellular junctions. Am J Pathol 2011;178(5):2424–36.
[132] Templin T, et al. Short-term effects of low-LET radiation on the endothelial barrier: uncoupling of PECAM-1 and the production of endothelial microparticles. Radiat Res 2016;186(6):602–13.

[133] Sharma P, Guida P, Grabham P. Effects of Fe particle irradiation on human endothelial barrier structure and function. Life Sci Space Res 2014;2:29–37.
[134] Kim KS, et al. Characterization of DNA damage-induced cellular senescence by ionizing radiation in endothelial cells. Int J Radiat Biol 2014;90(1):71–80.
[135] Zorkina YA, et al. Effect of gamma-irradiation on expression of tight and adherens junction protein mRNA on in vitro blood-brain barrier model. Bull Exp Biol Med 2014;158(1):127–36.
[136] Alabi RO, Farber G, Blobel CP. Intriguing roles for endothelial ADAM10/Notch signaling in the development of organ-specific vascular beds. Physiol Rev 2018; 98(4):2025–61.
[137] Shulz B, et al. ADAM10 regulates endothelial permeability and T-cell transmigration by proteolysis of vascular endothelial cadherin. Circ Res 2008;102:1192–201.
[138] Donners MM, et al. A disintegrin and metalloprotease 10 is a novel mediator of vascular endothelial growth factor-induced endothelial cell function in angiogenesis and is associated with atherosclerosis. Arterioscler Thromb Vasc Biol 2010;30(11): 2188–95.
[139] Mori K, et al. Mitogen-activated protein kinase, ERK1/2, is essential for the induction of vascular endothelial growth factor by ionizing radiation mediated by activator protein-1 in human glioblastoma cells. Free Radic Res 2000;33(2):157–66.
[140] Augustin HG, Kozian DH, Johnson RC. Differentiation of endothelial cells: analysis of the constitutive and activated endothelial cell phenotypes. Bioessays 1994;16(12): 901–6.
[141] Cines DB, et al. Endothelial cells in physiology and in the pathophysiology of vascular disorders. Blood 1998;91(10):3527–61.
[142] McCarthy SA, et al. Heterogeneity of the endothelial cell and its role in organ preference of tumour metastasis. Trends Pharmacol Sci 1991;12(12):462–7.
[143] Seok J, et al. Genomic responses in mouse models poorly mimic human inflammatory diseases. Proc Natl Acad Sci USA 2013;110(9):3507–12.
[144] Shay T, Lederer JA, Benoist C. Genomic responses to inflammation in mouse models mimic humans: we concur, apples to oranges comparisons won't do. Proc Natl Acad Sci USA 2015;112(4):E346.
[145] Takao K, Hagihara H, Miyakawa T. Reply to Warren et al. and Shay et al.: Commonalities across species do exist and are potentially important. Proc Natl Acad Sci USA 2015;112(4):E347–8.
[146] Takao K, Miyakawa T. Genomic responses in mouse models greatly mimic human inflammatory diseases. Proc Natl Acad Sci USA 2015;112(4):1167–72.
[147] Mendelsohn ME, Karas RH. Molecular and cellular basis of cardiovascular gender differences. Science 2005;308(5728):1583–7.
[148] Stewart FA, et al. Ionizing radiation accelerates the development of atherosclerotic lesions in ApoE-/- mice and predisposes to an inflammatory plaque phenotype prone to hemorrhage. Am J Pathol 2006;168(2):649–58.
[149] Wang Y, Boerma M, Zhou D. Ionizing radiation-induced endothelial cell senescence and cardiovascular diseases. Radiat Res 2016;186(2):153–61.
[150] Panganiban RA, Snow AL, Day RM. Mechanisms of radiation toxicity in transformed and non-transformed cells. Int J Mol Sci 2013;14(8):15931–58.
[151] Gueguen Y, Bontemps A, Ebrahimian TG. Adaptive responses to low doses of radiation or chemicals: their cellular and molecular mechanisms. Cell Mol Life Sci 2019;76(7):1255–73.
[152] Paris F, et al. Endothelial apoptosis as the primary lesion initiating intestinal radiation damage in mice. Science 2001;293(5528):293–7.
[153] Abderrahmani R, et al. PAI-1-dependent endothelial cell death determines severity of radiation-induced intestinal injury. PLoS One 2012;7(4):e35740.

[154] Rotolo J, et al. Anti-ceramide antibody prevents the radiation gastrointestinal syndrome in mice. J Clin Invest 2012;122(5):1786–90.
[155] Li YQ, et al. Endothelial apoptosis initiates acute blood-brain barrier disruption after ionizing radiation. Cancer Res 2003;63(18):5950–6.
[156] Shinohara C, et al. Apoptosis in the subependyma of young adult rats after single and fractionated doses of X-rays. Cancer Res 1997;57(13):2694–702.
[157] Ljubimova NV, et al. Endothelial cell population dynamics in rat brain after local irradiation. Br J Radiol 1991;64(766):934–40.
[158] Lu F, et al. Endothelial cells regulate p53-dependent apoptosis of neural progenitors after irradiation. Cell Death Dis 2012;3:e324.
[159] Mao XW, et al. Characterization of mouse ocular response to a 35-day spaceflight mission: Evidence of blood-retinal barrier disruption and ocular adaptations. Sci Rep 2019;9(1):8215.
[160] Tirodkar TS, Voelkel-Johnson C. Sphingolipids in apoptosis. Exp Oncol 2012;34(3): 231–42.
[161] Stancevic B, Kolesnick R. Ceramide-rich platforms in transmembrane signaling. FEBS Lett 2010;584(9):1728–40.
[162] Gulbins E. Regulation of death receptor signaling and apoptosis by ceramide. Pharmacol Res 2003;47(5):393–9.
[163] Kolesnick R, Fuks Z. Radiation and ceramide-induced apoptosis. Oncogene 2003; 22(37):5897–906.
[164] Zhang AY, et al. Acid sphingomyelinase and its redox amplification in formation of lipid raft redox signaling platforms in endothelial cells. Antioxidants Redox Signal 2007;9(7):817–28.
[165] Rizzieri KE, Hannun YA. Sphingolipid metabolism, apoptosis and resistance to cytotoxic agents: can we interfere? Drug Resist Updates 1998;1(6):359–76.
[166] Sun L, et al. Reactive oxygen species mediates 50-Hz magnetic field-induced EGF receptor clustering via acid sphingomyelinase activation. Int J Radiat Biol 2018; 94(7):678–84.
[167] Santana P, et al. Acid sphingomyelinase-deficient human lymphoblasts and mice are defective in radiation-induced apoptosis. Cell 1996;86(2):189–99.
[168] Hannun YA, Obeid LM. Ceramide: an intracellular signal for apoptosis. Trends Biochem Sci 1995;20(2):73–7.
[169] Obeid LM, Hannun YA. Ceramide: a stress signal and mediator of growth suppression and apoptosis. J Cell Biochem 1995;58(2):191–8.
[170] Jarvis WD, Grant S, Kolesnick RN. Ceramide and the induction of apoptosis. Clin Cancer Res 1996;2(1):1–6.
[171] Verheij M, et al. Requirement for ceramide-initiated SAPK/JNK signalling in stress-induced apoptosis. Nature 1996;380(6569):75–9.
[172] Marathe S, et al. Human vascular endothelial cells are a rich and regulatable source of secretory sphingomyelinase. Implications for early atherogenesis and ceramide-mediated cell signaling. J Biol Chem 1998;273(7):4081–8.
[173] Li X, et al. TRAIL death receptor 4 signaling via lysosome fusion and membrane raft clustering in coronary arterial endothelial cells: evidence from ASM knockout mice. J Mol Med 2013;91(1):25–36.
[174] Haimovitz-Friedman A, et al. Ionizing radiation acts on cellular membranes to generate ceramide and initiate apoptosis. J Exp Med 1994;180(2):525–35.
[175] Lin X, Fuks Z, Kolesnick R. Ceramide mediates radiation-induced death of endothelium. Crit Care Med 2000;28(4 Suppl. l):N87–93.
[176] Kolesnick RN, Haimovitz-Friedman A, Fuks Z. The sphingomyelin signal transduction pathway mediates apoptosis for tumor necrosis factor, fas, and ionizing radiation. Biochem Cell Biol 1994;72(11–12):471–4.

[177] Liao WC, et al. Ataxia telangiectasia-mutated gene product inhibits DNA damage-induced apoptosis via ceramide synthase. J Biol Chem 1999;274(25):17908–17.
[178] Garcia-Barros M, et al. Tumor response to radiotherapy regulated by endothelial cell apoptosis. Science 2003;300(5622):1155–9.
[179] Zhu H, et al. The effects of ASMase mediated endothelial cell apoptosis in multiple hypofractionated irradiations in CT26 tumor bearing mice. Asian Pac J Cancer Prev 2015;16(11):4543–8.
[180] Lee CL, et al. p53 functions in endothelial cells to prevent radiation-induced myocardial injury in mice. Sci Signal 2012;5(234):ra52.
[181] Panganiban RA, Mungunsukh O, Day RM. X-irradiation induces ER stress, apoptosis, and senescence in pulmonary artery endothelial cells. Int J Radiat Biol 2013;89(8):656–67.
[182] Ferreira FU, et al. Endothelial cells tissue-specific origins affects their responsiveness to TGF-β2 during endothelial-to-mesenchymal transition. Int J Mol Sci 2019;20:458.
[183] Schwarz MA, et al. Emerging pulmonary vasculature lacks fate specification. Am J Physiol Lung Cell Mol Physiol 2009;296(1):L71–81.
[184] Aureli M, et al. Exploring the link between ceramide and ionizing radiation. Glycoconj J 2014;31(6–7):449–59.
[185] Hayflick L, Moorhead PS. The serial cultivation of human diploid cell strains. Exp Cell Res 1961;25:585–621.
[186] de Magalhaes JP, Passos JF. Stress, cell senescence and organismal ageing. Mech Ageing Dev 2018;170:2–9.
[187] Muller M. Cellular senescence: molecular mechanisms, in vivo significance, and redox considerations. Antioxidants Redox Signal 2009;11(1):59–98.
[188] van Deursen JM. The role of senescent cells in ageing. Nature 2014;509(7501):439–46.
[189] Tchkonia T, et al. Cellular senescence and the senescent secretory phenotype: therapeutic opportunities. J Clin Invest 2013;123(3):966–72.
[190] Dong X, et al. NEMO modulates radiation-induced endothelial senescence of human umbilical veins through NF-kappaB signal pathway. Radiat Res 2015;183(1):82–93.
[191] Ungvari Z, et al. Ionizing radiation promotes the acquisition of a senescence-associated secretory phenotype and impairs angiogenic capacity in cerebromicrovascular endothelial cells: role of increased DNA damage and decreased DNA repair capacity in microvascular radiosensitivity. J Gerontol A Biol Sci Med Sci 2013;68(12):1443–57.
[192] Panganiban RA, Day RM. Inhibition of IGF-1R prevents ionizing radiation-induced primary endothelial cell senescence. PLoS One 2013;8(10):e78589.
[193] Yentrapalli R, et al. The PI3K/Akt/mTOR pathway is implicated in the premature senescence of primary human endothelial cells exposed to chronic radiation. PLoS One 2013;8(8):e70024.
[194] Yentrapalli R, et al. Quantitative proteomic analysis reveals induction of premature senescence in human umbilical vein endothelial cells exposed to chronic low-dose rate gamma radiation. Proteomics 2013;13(7):1096–107.
[195] Siegel JA, et al. The BEIR VII estimates of low-dose radiation health risks are based on faulty assumptions and data analyses: a call for reassessment. J Nucl Med 2018;59(7):1017–9.
[196] Duncan JR, et al. Reply: radiation dose does matter: mechanistic insights into DNA damage and repair support the linear no-threshold model of low-dose radiation health risks. J Nucl Med 2019;60(3):437–8.

[197] Lafargue A, et al. Ionizing radiation induces long-term senescence in endothelial cells through mitochondrial respiratory complex II dysfunction and superoxide generation. Free Radic Biol Med 2017;108:750–9.
[198] Vucic D. XIAP at the crossroads of cell death and inflammation. Oncotarget 2018; 9(44):27319–20.
[199] Liston P, et al. Identification of XAF1 as an antagonist of XIAP anti-caspase activity. Nat Cell Biol 2001;3(2):128–33.
[200] Gala MK, et al. Germline mutations in oncogene-induced senescence pathways are associated with multiple sessile serrated adenomas. Gastroenterology 2014;146(2): 520–9.
[201] Heo JI, et al. XIAP-associating factor 1, a transcriptional target of BRD7, contributes to endothelial cell senescence. Oncotarget 2016;7(5):5118–30.
[202] Kumar M, Seeger W, Voswinckel R. Senescence-associated secretory phenotype and its possible role in chronic obstructive pulmonary disease. Am J Respir Cell Mol Biol 2014;51(3):323–33.
[203] Salminen A, Kauppinen A, Kaarniranta K. Emerging role of NF-kappaB signaling in the induction of senescence-associated secretory phenotype (SASP). Cell Signal 2012; 24(4):835–45.
[204] Wu ZH, et al. Molecular linkage between the kinase ATM and NF-kappaB signaling in response to genotoxic stimuli. Science 2006;311(5764):1141–6.
[205] Galluzzi L, et al. No death without life: vital functions of apoptotic effectors. Cell Death Differ 2008;15(7):1113–23.
[206] Imaizumi N, et al. Radiotherapy suppresses angiogenesis in mice through TGF-betaRI/ALK5-dependent inhibition of endothelial cell sprouting. PLoS One 2010;5(6):e11084.
[207] Bostrom M, et al. A role for endothelial cells in radiation-induced inflammation. Int J Radiat Biol 2018;94(3):259–71.
[208] Bostrom M, et al. Irradiation to the young mouse brain caused long-term, progressive depletion of neurogenesis but did not disrupt the neurovascular niche. J Cerebr Blood Flow Metabol 2013;33(6):935–43.
[209] De Pascalis I, et al. Endothelial trans-differentiation in glioblastoma recurring after radiotherapy. Mod Pathol 2018;31(9):1361–6.
[210] Sandor N, et al. Low dose cranial irradiation-induced cerebrovascular damage is reversible in mice. PLoS One 2014;9(11):e112397.
[211] McRobb LS, et al. Ionizing radiation reduces ADAM10 expression in brain microvascular endothelial cells undergoing stress-induced senescence. Aging 2017;9(4): 1248–68.
[212] Si S, et al. Hematopoietic insults damage bone marrow niche by activating p53 in vascular endothelial cells. Exp Hematol 2018;63:41–51 e1.
[213] Azimzadeh O, et al. Integrative proteomics and targeted transcriptomics analyses in cardiac endothelial cells unravel mechanisms of long-term radiation-induced vascular dysfunction. J Proteome Res 2015;14(2):1203–19.
[214] Pinte S, et al. Endothelial cell activation is regulated by epidermal growth factor-like domain 7 (Egfl7) during inflammation. J Biol Chem 2016;291(46):24017–28.
[215] Day RM, Snow AL, Panganiban RA. Radiation-induced accelerated senescence: a fate worse than death? Cell Cycle 2014;13(13):2011–2.
[216] Holler V, et al. Pravastatin limits radiation-induced vascular dysfunction in the skin. J Invest Dermatol 2009;129(5):1280–91.
[217] Quarmby S, Kumar P, Kumar S. Radiation-induced normal tissue injury: role of adhesion molecules in leukocyte-endothelial cell interactions. Int J Cancer 1999; 82(3):385–95.

[218] Halle M, et al. Endothelial activation with prothrombotic response in irradiated microvascular recipient veins. J Plast Reconstr Aesthetic Surg 2010;63(11):1910–6.
[219] Surh YJ, et al. Redox-sensitive transcription factors as prime targets for chemoprevention with anti-inflammatory and antioxidative phytochemicals. J Nutr 2005; 135(12 Suppl. l):2993S–3001S.
[220] Kim I, et al. Vascular endothelial growth factor expression of intercellular adhesion molecule 1 (ICAM-1), vascular cell adhesion molecule 1 (VCAM-1), and E-selectin through nuclear factor-kappa B activation in endothelial cells. J Biol Chem 2001;276(10):7614–20.
[221] Austin SA, Sens MA, Combs CK. Amyloid precursor protein mediates a tyrosine kinase-dependent activation response in endothelial cells. J Neurosci 2009;29(46): 14451–62.
[222] Woodruff KH, et al. Morphologic and morphometric analysis of the early effects of x-ray and heavy-ion irradiation of hamster lung. Am J Pathol 1976;82(2):287–98.
[223] Hebard DW, Jackson KL, Christensen GM. The chronological development of late radiation injury in the liver of the rat. Radiat Res 1980;81(3):441–54.
[224] Robbins ME, et al. Sequential evaluation of radiation-induced glomerular ultrastructural changes in the pig kidney. Radiat Res 1993;135(3):351–64.
[225] Herskind C, Bamberg M, Rodemann HP. The role of cytokines in the development of normal-tissue reactions after radiotherapy. Strahlenther Onkol 1998;174(Suppl. 3): 12–5.
[226] Rodemann H-P. Role of radiation-induced signaling proteins in the response of the vascular and connective tissues. In: Nieder C, Milas L, Ang KK, editors. Modification of radiation response. Cytokines, growth factors, and other biological targets. Berlin: Springer; 2003.
[227] Cheki M, et al. The radioprotective effect of metformin against cytotoxicity and genotoxicity induced by ionizing radiation in cultured human blood lymphocytes. Mutat Res 2016;809:24–32.
[228] Ossetrova NI, et al. Non-human primate total-body irradiation model with limited and full medical supportive care including filgrastim for biodosimetry and injury assessment. Radiat Protect Dosim 2016;172(1–3):174–91.
[229] Ossetrova NI, et al. Early-response biomarkers for assessment of radiation exposure in a mouse total-body irradiation model. Health Phys 2014;106(6):772–86.
[230] Ossetrova NI, et al. Acute radiation syndrome severity score system in mouse total-body irradiation model. Health Phys 2016;111(2):134–44.
[231] Ossetrova NI, et al. Combined approach of hematological biomarkers and plasma protein SAA for improvement of radiation dose assessment triage in biodosimetry applications. Health Phys 2010;98(2):204–8.
[232] Ossetrova NI, et al. Biomarkers for radiation biodosimetry and injury assessment after mixed-field (neutron and gamma) radiation in the mouse total-body irradiation model. Health Phys 2018;115(6):727–42.
[233] McCart EA, et al. Delayed captopril administration mitigates hematopoietic injury in a murine model of total body irradiation. Sci Rep 2019;9(1):2198.
[234] Calveley VL, et al. Partial volume rat lung irradiation: temporal fluctuations of in-field and out-of-field DNA damage and inflammatory cytokines following irradiation. Int J Radiat Biol 2005;81(12):887–99.
[235] Kunder CA, St John AL, Abraham SN. Mast cell modulation of the vascular and lymphatic endothelium. Blood 2011;118(20):5383–93.
[236] Krystel-Whittemore M, Dileepan KN, Wood JG. Mast cell: a multi-functional master cell. Front Immunol 2015;6:620.
[237] Aldenborg F, et al. Mast cells and biogenic amines in radiation-induced pulmonary fibrosis. Am J Respir Cell Mol Biol 1993;8(1):112–7.

[238] Canney PA, Dean S. Transforming growth factor beta: a promotor of late connective tissue injury following radiotherapy? Br J Radiol 1990;63(752):620–3.
[239] Zheng H, Wang J, Hauer-Jensen M. Role of mast cells in early and delayed radiation injury in rat intestine. Radiat Res 2000;153(5 Pt 1):533–9.
[240] Albrecht M, et al. Ionizing radiation induces degranulation of human mast cells and release of tryptase. Int J Radiat Biol 2007;83(8):535–41.
[241] Park KR, et al. Mast cells contribute to radiation-induced vascular hyperpermeability. Radiat Res 2016;185(2):182–9.
[242] Sylven B. Studies on the liberation of sulphuric acids from the granules of the mast cells in the subcutaneous connective tissue after exposure to roentgen and gamma rays. Acta Radiol 1940;21:206.
[243] Klimenko NA, Pavlova EA. [Mast cell reaction to total irradiation]. Radiats Biol Radioecol 1997;37(3):395–8.
[244] van den Brenk HAS. Observations on mast cell changes, histamine release and local tissue damage in rats following x-irradiation. Br J Exp Pathol 1958;39:356–66.
[245] Smith DE, Lewis YS. Effects of total body X-irradiation on the tissue mast cells. Proc Soc Exp Biol Med 1953;82:208–12.
[246] Lasser EC, Stenstrom KW. Elevation of circulating blood histamine in patients undergoing deep roentgen therapy. Am J Roentgenol Radium Ther Nucl Med 1954;72:985–8.
[247] Davis J. Antihistamines in allergic and non-allergic dermatoses. Bull N Y Acad Med 1950;26:686–99.
[248] Mains PM. Anit-histamine ointments for skin protection in radiation therapy. Radiology 1949;52:579–88.
[249] Mount D, Bruce WR. Local plasma volume and vascular permeability of rabbit skin after radiation. Radiat Res 1964;23:430–45.
[250] Willoughby DA. Pharmacological aspects of the vascular permeability changes in the rat's intestine following abdominal radiation. Br J Radiol 1960;33:515–9.
[251] Rubin DB. The radiation biology of the vascular endothelium. Boca Raton: CRC Press; 1997.
[252] Riekki R, et al. The production of collagen and the activity of mast-cell chymase increase in human skin after irradiation therapy. Exp Dermatol 2004;13(6):364–71.
[253] Abbott NJ, Ronnback L, Hansson E. Astrocyte-endothelial interactions at the blood-brain barrier. Nat Rev Neurosci 2006;7(1):41–53.
[254] Hawkins BT, Davis TP. The blood-brain barrier/neurovascular unit in health and disease. Pharmacol Rev 2005;57(2):173–85.
[255] Liddelow SA, Barres BA. Reactive astrocytes: production, function, and therapeutic potential. Immunity 2017;46(6):957–67.
[256] Reichenbach A, Wolburg H. In: Dettemann H, Ransom BR, editors. Neuroglia. 2nd ed. New York: Oxford University Press; 2004.
[257] McCaffrey G, et al. Occludin oligomeric assembly at tight junctions of the blood-brain barrier is disrupted by peripheral inflammatory hyperalgesia. J Neurochem 2008;106(6):2395–409.
[258] Gutierrez Y, et al. Sibling astrocytes share preferential coupling via gap junctions. Glia 2019;67(10):1852–8.
[259] Huber VJ, et al. Aquaporin-4 facilitator TGN-073 promotes interstitial fluid circulation within the blood-brain barrier: [17O]H_2O JJVCPE MRI study. Neuroreport 2018;29(9):697–703.
[260] Cheng X, et al. Morphological and functional alterations of astrocytes responding to traumatic brain injury. J Integr Neurosci 2019;18(2):203–15.
[261] Nishioku T, et al. Tumor necrosis factor-alpha mediates the blood-brain barrier dysfunction induced by activated microglia in mouse brain microvascular endothelial cells. J Pharmacol Sci 2010;112(2):251–4.

[262] Brioschi S, Colonna M. The CNS immune-privilege goes down the drain(age). Trends Pharmacol Sci 2019;40(1):1–3.

[263] Pekny M, Wilhelmsson U, Pekna M. The dual role of astrocyte activation and reactive gliosis. Neurosci Lett 2014;565:30–8.

[264] Escartin C, Bonvento G. Targeted activation of astrocytes: a potential neuroprotective strategy. Mol Neurobiol 2008;38(3):231–41.

[265] Schiweck J, Eickholt BJ, Murk K. Important shapeshifter: mechanisms allowing astrocytes to respond to the changing nervous system during development, injury and disease. Front Cell Neurosci 2018;12:261.

[266] Uyeda CT, Eng LF, Bignami A. Immunological study of the glial fibrillary acidic protein. Brain Res 1972;37(1):81–9.

[267] Bylicky MA, Mueller GP, Day RM. Mechanisms of endogenous neuroprotective effects of astrocytes in brain injury. Oxid Med Cell Longev 2018;2018:6501031.

[268] Bylicky MA, Mueller GP, Day RM. Radiation resistance of normal human astrocytes: the role of non-homologous end joining DNA repair activity. J Radiat Res 2019;60(1):37–50.

[269] Hwang SY, et al. Ionizing radiation induces astrocyte gliosis through microglia activation. Neurobiol Dis 2006;21(3):457–67.

[270] Balentova S, et al. Effect of whole-brain irradiation on the specific brain regions in a rat model: metabolic and histopathological changes. Neurotoxicology 2017;60: 70–81.

[271] Shigemoto-Mogami Y, Hoshikawa K, Sato K. Activated microglia disrupt the blood-brain barrier and induce chemokines and cytokines in a rat in vitro model. Front Cell Neurosci 2018;12:494.

[272] Wilson CM, et al. Radiation-induced astrogliosis and blood-brain barrier damage can be abrogated using anti-TNF treatment. Int J Radiat Oncol Biol Phys 2009;74(3): 934–41.

[273] Evstatiev R, Gasche C. Iron sensing and signalling. Gut 2012;61(6):933–52.

[274] Daher R, Karim Z. Iron metabolism: state of the art. Transfus Clin Biol 2017;24(3): 115–9.

[275] Andrews NC. Iron homeostasis: insights from genetics and animal models. Nat Rev Genet 2000;1(3):208–17.

[276] Nairz M, et al. "Pumping iron"-how macrophages handle iron at the systemic, microenvironmental, and cellular levels. Pflügers Archiv 2017;469(3–4):397–418.

[277] Winterbourn CC. Toxicity of iron and hydrogen peroxide: the Fenton reaction. Toxicol Lett 1995;82–83:969–74.

[278] Valko M, et al. Redox- and non-redox-metal-induced formation of free radicals and their role in human disease. Arch Toxicol 2016;90(1):1–37.

[279] Carocci A, et al. Oxidative stress and neurodegeneration: the involvement of iron. Biometals 2018;31(5):715–35.

[280] Isidori A, et al. Iron toxicity - Its effect on the bone marrow. Blood Rev 2018;32(6): 473–9.

[281] Sullivan JL. Iron in arterial plaque: modifiable risk factor for atherosclerosis. Biochim Biophys Acta 2009;1790(7):718–23.

[282] Chai X, et al. ROS-mediated iron overload injures the hematopoiesis of bone marrow by damaging hematopoietic stem/progenitor cells in mice. Sci Rep 2015; 5:10181.

[283] Nairz M, et al. Iron at the interface of immunity and infection. Front Pharmacol 2014;5:152.

[284] Meder J, Michalowski A. Changes in cellularity and/or weight of mouse hemopoietic tissues as a measure of acute radiation effects. Arch Immunol Ther Exp 1980;28(1): 9–18.

[285] Shao L, Luo Y, Zhou D. Hematopoietic stem cell injury induced by ionizing radiation. Antioxidants Redox Signal 2014;20(9):1447–62.
[286] Peslak SA, et al. Sublethal radiation injury uncovers a functional transition during erythroid maturation. Exp Hematol 2011;39(4):434–45.
[287] Puchala M, Szweda-Lewandowska Z, Kiefer J. The influence of radiation quality on radiation-induced hemolysis and hemoglobin oxidation of human erythrocytes. J Radiat Res 2004;45(2):275–9.
[288] Zhang B, et al. Erythrocyte stiffness during morphological remodeling induced by carbon ion radiation. PLoS One 2014;9(11):e112624.
[289] Barasch J, et al. Disposal of iron by a mutant form of lipocalin 2. Nat Commun 2016; 7:12973.
[290] Xie L-H, et al. Mechanisms of an increased level of serum iron in gamma-irradiated mice. Radiat Environ Biophys 2016;55:81–8.
[291] Zhang XH, et al. Development of serum iron as a biological dosimeter in mice. Radiat Res 2013;179(6):684–9.
[292] Zhang J, et al. Lowering iron levels protects against bone loss in focally irradiated and contralateral femurs through distinct mechanisms. Bone 2019;120:50–60.
[293] Carlini RG, et al. Apoptotic stress pathway activation mediated by iron on endothelial cells in vitro. Nephrol Dial Transplant 2006;21(11):3055–61.
[294] Chan S, et al. Deferiprone inhibits iron overload-induced tissue factor bearing endothelial microparticle generation by inhibition oxidative stress induced mitochondrial injury, and apoptosis. Toxicol Appl Pharmacol 2018;338:148–58.
[295] Imai T, et al. Intracellular Fe(2+) accumulation in endothelial cells and pericytes induces blood-brain barrier dysfunction in secondary brain injury after brain hemorrhage. Sci Rep 2019;9(1):6228.
[296] Balla J, et al. Haem, haem oxygenase and ferritin in vascular endothelial cell injury. Nephrol Dial Transplant 2003;18(Suppl. 5):v8–12.
[297] Fogg S, et al. Iron regulates hyperoxia-dependent human heme oxygenase 1 gene expression in pulmonary endothelial cells. Am J Respir Cell Mol Biol 1999;20(4): 797–804.
[298] Gall T, Balla G, Balla J. Heme, heme oxygenase, and endoplasmic reticulum stress-a new insight into the pathophysiology of vascular diseases. Int J Mol Sci 2019;20(15).
[299] Burdo JR, et al. Regulation of the profile of iron-management proteins in brain microvasculature. J Cerebr Blood Flow Metabol 2004;24(1):67–74.
[300] Daglas M, Adlard PA. The involvement of iron in traumatic brain injury and neurodegenerative disease. Front Neurosci 2018;12:981.
[301] Won SM, et al. Iron mediates endothelial cell damage and blood-brain barrier opening in the hippocampus after transient forebrain ischemia in rats. Exp Mol Med 2011; 43(2):121–8.
[302] Kondo Y, et al. Regional differences in late-onset iron deposition, ferritin, transferrin, astrocyte proliferation, and microglial activation after transient forebrain ischemia in rat brain. J Cerebr Blood Flow Metabol 1995;15(2):216–26.
[303] Wang HK, et al. Free radical production in CA1 neurons induces MIP-1alpha expression, microglia recruitment, and delayed neuronal death after transient forebrain ischemia. J Neurosci 2008;28(7):1721–7.
[304] Baskar R, et al. Cancer and radiation therapy: current advances and future directions. Int J Med Sci 2012;9(3):193–9.
[305] Gale RP. Medical and policy considerations for nuclear and radiation accidents, incidents and terrorism. Curr Opin Hematol 2017;24(6):496–501.
[306] Gale RP, Armitage JO, Hashmi SK. Emergency response to radiological and nuclear accidents and incidents. Br J Haematol 2019. https://doi.org/10.1111/bjh.16138. In press.

[307] Mukherjee S, Chakraborty A. Radiation-induced bystander phenomenon: insight and implications in radiotherapy. Int J Radiat Biol 2019;95(3):243–63.
[308] Mukherjee S, et al. Impact of inflammatory signaling on radiation biodosimetry: mouse model of inflammatory bowel disease. BMC Genom 2019;20(1):329.
[309] Gaugler MH, et al. Intestinal epithelial cell dysfunction is mediated by an endothelial-specific radiation-induced bystander effect. Radiat Res 2007;167(2):185–93.

CHAPTER THREE

How severe RNA virus infections such as SARS-CoV-2 disrupt tissue and organ barriers—Reconstitution by mesenchymal stem cell-derived exosomes

Jason Sanders, MD [1], E. Marion Schneider[2]
[1]Chief Medical Officer, EV Biologics, Inc., Wyoming, United States
[2]Division of Experimental Anaesthesiology, University Hospital Ulm, Ulm, Germany

1. COVID-19 and the contribution by immune effector cells

1.1 SARS-COV-2

SARS-Cov-2 is a coronavirus with a single-strand, positive-sense RNA genome with multiple open reading frames that codes for 10 proteins including the nucleocapsid and the Spike protein, which is important for infection [1]. SARS-Cov-2 primarily infects type II alveolar epithelial cells (AECs) because they have a high concentration of ACE2 receptors, which have a strong interaction with the viral Spike protein, facilitating internalization of the virus [2,3]. COVID-19, the viral respiratory illness that results from SARS-Cov-2 infection, initially presents with mild symptoms for several days concurrent with the highest levels of viral shedding [4]. The inflammatory damage of COVID-19 follows as the natural immune response to the virus causes the release of high levels of inflammatory mediators, such as tumor necrosis factor-α (TNF-α) and interleukin-6 (IL-6), in a sustained pattern distinct from bacterial sepsis or influenza [5,6]. The rapid clinical deterioration about 7 days after initial onset of symptoms suggests that the respiratory failure in COVID-19 results from a unique pattern of immune dysregulation characterized by macrophage activation syndrome (MAS) or profound depletion of CD4 lymphocytes, CD19 lymphocytes, and natural killer (NK) cells. The persistent immune response, despite falling

Tissue Barriers in Disease, Injury and Regeneration
ISBN: 978-0-12-818561-2
https://doi.org/10.1016/B978-0-12-818561-2.00004-7

viral titers in this inflammatory phase, leads to progressive organ and tissue injury, suggesting that the immune-mediated damage is more significant than the viral cytopathic damage [7,8].

2. Immune-mediated pathogenesis

SARS-Cov-2 induces immune dysregulation in a different pattern than influenza or SARS-CoV. Chemokines (CCL2 [MCP 1], CCL3 [MIP1α], CXCL1, CXCL5, and CXCL10) are significantly upregulated in COVID-19 and act to recruit macrophages, neutrophils, and effector T cells [9]. These inflammatory conditions are linked to defective antigen presentation. Individual viral antigens may be responsible for lymphocyte apoptosis which further drives monocytes to produce high levels of TNF-α and IL-6. TNF-α activates endothelial cells to recruit leukocytes, stimulates neutrophils, and increases the levels NF-κB, AP-1, IL-8, and caspases, which induce apoptosis of target tissues [10]. The ratio of lymphocytes to neutrophils appears to be an important predictor for outcome of COVID-19. Finally, advanced cases of COVID-19 respiratory failure are characterized by features of immune dysregulation or MAS.

Over-production of IL-6 promotes immune dysregulation with inhibition of HLA-DR expression on CD14 monocytes contributing to the impaired T cell response. IFN-γ produced by CD4+ TH1, NK, and CD8+ T cells normally increases MHC class I and II to promote killing of infected cells by T cells, increased macrophage activity and production of IgG antibodies, but there are very low levels of IFN-γ detected in COVID-19. Immune dysregulation caused by COVID-19 features lower counts of CD4+ T cells, CD8+ T cells, and NK cells than at the intermediate immune state. Recently, SARS-CoV2 specific genes have been identified which counter-regulate appropriate interferon responses in COVID-19. Infection by SARS-CoV-2 is characterized by fewer CD4+ T cells but more NK cells and B cells than H1N1 influenza. IL-17 production, indicating Th17 function, is downregulated in COVID-19 patients with immune dysregulation. There is also some evidence to suggest that SARS-Cov-2 infects and induces apoptosis of T lymphocytes, such as CD4+ TH1 cells that activate macrophages, CD4+ TH17 cells that activate neutrophils, and CD8+ cytotoxic T cells that kill infected cells. NK cells that kill infected or damaged cells are also reduced in SARS-CoV and SARS-CoV2 infection possibly resulting from the rapidly propagating virus [11]. In some cases of COVID-19, elevated levels of IL-1β promote a MAS with a pattern that may be similar to secondary hemophagocytic lymphohistiocytosis [12,13].

3. Acute lung injury, increased endothelial permeability, and loss of organ barrier function

The massive release of inflammatory mediators termed cytokine storm can cause an acute lung injury (ALI) characterized by disruption of junctions between cells, damage to AECs, damage to pulmonary capillary endothelial cells, and loss of alveolar fluid clearance mechanism [14,15]. Inflammatory mediators (cytokines and chemokines) released by type II AECs increase vasodilation, leukocyte adhesion, and capillary permeability [16,17]. Proteolytic enzymes and reactive oxygen species released by neutrophils and macrophages within the alveoli damage alveolar cells and the extracellular matrix [18,19]. Disruption of the junctions between the cells in the alveolar-capillary barrier promotes recruitment of nonspecific immune cells (e.g., neutrophils, macrophages) and allows exudative leak from the pulmonary capillaries [20]. In this context, the detrimental role of the kallikrein-bradykinin system is likely responsible for extensive fluid loss and inflammation for endothelial barriers. Neutrophils, macrophages, and other immune cells thus evade from the circulation, accumulate in the lungs, and increase their release of inflammatory mediators creating a positive feedback loop.

Viral inhibition of ACE2-mediated inactivation of des-Arg bradykinin and proinflammatory cytokine-mediated upregulation of bradykinin receptor type 1 (B1) on endothelial cells may cause pulmonary angioedema consistent with the radiographic findings of COVID-19 [21], which is in part because of increased expression of bradykinin receptor 2 (B2). The diffuse alveolar damage and angioedema could progress to acute respiratory distress syndrome with accumulation of proteinaceous fluid in alveoli that interferes with arterial oxygenation. This proteinous fluid consists of gelatinous hyaluronic acid and explains the opaque signs of COVID-19 lung images (Fig. 3.1). Diffuse thrombosis in the pulmonary vasculature and ventilator pressure support may also increase pulmonary artery pressure causing pulmonary hypertension, which could have secondary cardiovascular effects.

4. Endogenous repair systems

Endogenous repair systems that regulate the immune response and stimulate tissue regeneration exist to promote recovery from ALI but may

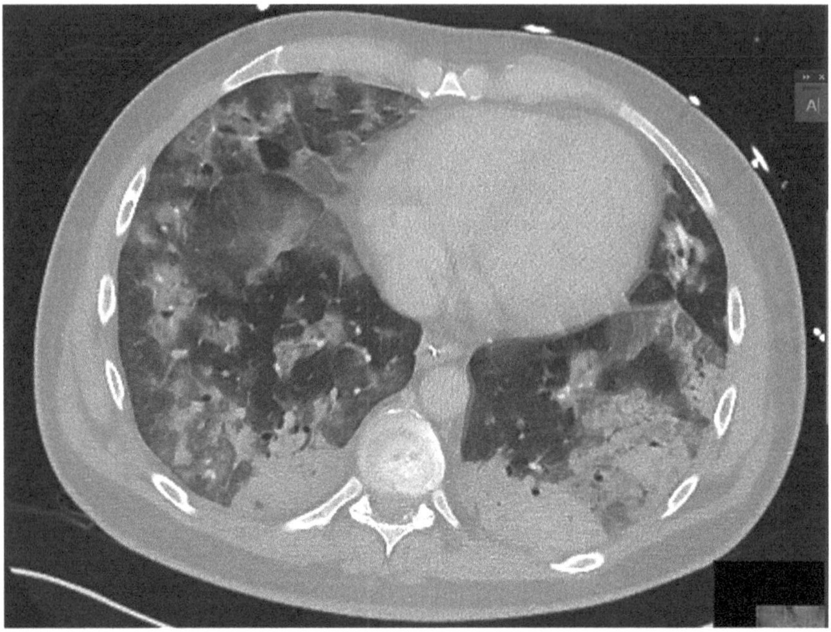

Figure 3.1 Mechanisms of action of mesenchymal stem cell-derived exosomes in acute respiratory distress syndrome [70].

be overwhelmed by the cytokine storm of COVID-19. The natural regulatory mechanisms of the immune system include production of antiinflammatory cytokines such as IL-10, TGF-β3, and IL-1ra, as well as immune cells with more regulatory phenotypes such as M2 macrophages and regulatory T cells [22]. Antiinflammatory cytokines (IL-10) promote phenotypes by polarizing M1 macrophages, which release high levels of inflammatory mediators, into tissue regenerative M2 macrophages, which release much lower levels of inflammatory mediators and also improve phagocytosis of cell debris, and danger ligands [23—25]. Antiinflammatory cytokines (TGF-β3) promote polarization of T lymphocytes from TH1, which produce high levels of inflammatory mediators to Treg cells, which regulate the cytotoxic activity of cytotoxic T lymphocytes (CTLs) and NK cells [26]. Regenerative cytokines including Ang-1, KGF, HGF, VEGF, EGF, Tsp1, S1P promote repair of the junctions between cells in the alveolar-capillary barrier, repair of the damaged alveoli and bronchioles by type II AECs and bronchoalveolar stem cells, and repair of pulmonary capillaries by epithelial cells [27—30].

5. The role of mesenchymal stem cells

In newborns and young children, mesenchymal stem cells (MSCs) are found in almost every organ and tissue. In adults, these large amounts of MSCs have been shown to be restricted to fat tissues including the bone marrow (BM). In addition to their potential to function as progenitors, MSCs have demonstrated the capacity to modulate the immune response and promote tissue regenerat ion. MSCs exert their paracrine effects through the release of soluble mediators including anti-inflammatory and regenerative cytokines and extracellular vesicles containing transcriptionally active RNA species, such as miRNA, lncRNA, and effector species. These soluble mediators and RNA species could potentially arrest the inflammatory response, repair the damage to the alveoli and pulmonary capillaries, and allow the immune system to clear the virus [20,30–36]. A Chinese study at Shanghai University on COVID-19 treatment using a master cell bank created from stromal progenitor cells harvested from lipoaspirate successfully treated seven patients who recovered, while three patients treated with placebo progressed to severe viral pneumonia or death [37]. An Israeli study of COVID-19 treatment using a clonal cell line of expanded, placental, MSCs treated six critically ill COVID-19 patients under a compassionate use program and demonstrated improvement in respiratory parameters and a 100% survival rate [38]. In the Israel study, four patients had other organ failures (cardiovascular, renal failure) and the placental MSCs therapy promoted recovery of the other organ failures as well as improvement in unrelated pre-existing conditions. Nearly 1000 clinical trials investigating administration of MSCs derived from sources such as BM, adipose tissue, cord blood, and others have also demonstrated a well-established safety profile of MSCs [39].

6. Extracellular vesicles: Exosomes and small microvesicles

MSCs exert the vast majority of their paracrine effects through the release of EVs, vesicles of roughly 50–1000 nm in diameter that are secreted by all cell types [40]. Small EVs (sEVs, 30–200 nm diameter, (Fig. 3.2), harvested using different protocols from cell culture supernatants of MSCs grown under diverse culture conditions, have been reported to be therapeutically active in various preclinical models [41]. EVs further may be responsible for the tissue regenerative effects observed with atopically transplanted

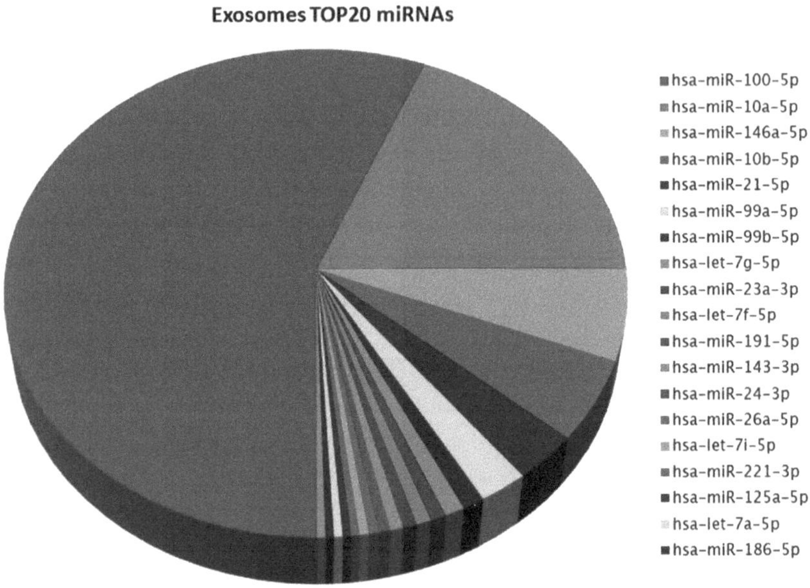

Figure 3.2 Expression of miRNA in purified mesenchymal stem cell-derived exosomes (small extracellular vesicles).

MSCs in various animal models. The origin of MSC-sEVs, a population of exosomes and small microvesicles produced by MSCs, suggests a similar safety profile to their parent cells that could expedite the development of clinical applications.

MSCs-EVs are internalized by target cells through endocytosis, and delivery of RNA species influences the behavior of these cells. Noncoding RNA such as miRNA, abundant in MSC-sEVs, target specific processes and pathways within target cells. The RNA content of MSC-sEVs confers their biological properties, which include being anti-inflammatory (e.g., miR1, miR100, miR181c), immunomodulatory (e.g., miR146a), proangiogenic (e.g., let-7, miR29), prosynthetic (e.g., miR92-3p, miR140-5p), antiapoptotic (e.g., miR21, miR199a), antifibrotic (e.g., miR21, miR23a, miR125b), and tumor suppressive (e.g., miR15a, miR145) (Fig. 3.2). It is important to note that the biological effects of MSC-sEVs are not the result of any single factor, but rather the combined activity of the abundant RNA moleculesaffecting multiple cellular pathways in the context of target cell behavior prior to internalization.

7. Tissue reconstitutive mechanisms by mesenchymal stem cell-small extracellular vesicles in COVID-19

Internalization of MSC-sEVs by cells exhibiting pathologic behavior may influence cellular pathways to regulate the hyperinflammatory response to SARS-CoV-2 and promote regeneration of damaged pulmonary and other tissues. MSC-sEVs may positively regulate expression of proteins such as TGF-β3, IL-10, IL-4, TNFR, IL-1ra, and HGF that could modulate the excessive immune response that contributes to tissue damage in COVID-19 [42—53]. MSC-sEVs could also promote alveolar tissue regeneration through Enhanced expression of angiopoietin-1 after endocytosis of MSC-sEVs could also promote alveolar tissue regeneration and decrease alveolar-capillary barrier permeability in ALI [28,29]. MSC-sEV induced expression of KGF (FGF-7) may also contribute to repair of AECs and increased alveolar fluid clearance [27]. EGF and HGF expression induced by MSC-sEVs has been demonstrated to be mitogenic for type II AECs and promoted alveolar regeneration [54]. One small cohort study in humans demonstrated direct evidence of a therapeutic benefit of MSC-sEVs in the treatment of COVID-19 with significant improvements in absolute neutrophil count, lymphopenia and acute phase reactants including C-reactive protein, ferritin, and D-dimer. Another phase I study of exosomes bearing CD24 also demonstrated complete recovery in all 30 patients with moderate to severe COVID-19, and recovery within three to five days in 29 of these patients.()

Transfer of MSC-sEV miRNA has also demonstrated beneficial effects on pathologic cellular processes that are believed to contribute to the pathogenesis of COVID-19. miR-455-3p inhibited the activation and cytokine production of macrophages challenged with lipopolysaccharide (LPS) both in vivo and in vitro and reduced levels of IL-6, G-CSF, IL-17, IL-10, IP-10 (CXCL10), MCP-1 (CCL2) [55]. miR-146a-5p targeted the expression of IRAK-1 and TRAF-6, significantly suppressed LPS-mediated TNF-α, IL-6, and IL-1β induction in alveolar macrophages, and increased IL-10, M2 macrophage polarization, and phagocytosis [56—60]. miR-223 targeted the transcription factor Pknox1 decreasing IL-1β, TNF-α, IL-6, and NF-κB, suppressed the proinflammatory activation of macrophages and promoted the alternative antiinflammatory M2 phenotype [61,62]. miR-511-3p targeted Rho-associated coiled-coil containing protein kinase 2 (Rock2), which is a serine threonine kinase that phosphorylates IRF4, and thus promoted the expression of M2-related genes [63]Fig. 3.2, MiR-100, an mTOR inhibitor, positively regulated autophagy, attenuated bleomycin-induced cellular apoptosis in type II AECs and reduces the levels of proinflammatory cytokines IL-6, IL-8, and TNF-α [64]. miR-30b-3p and miRNA-21-5p reduced apoptosis of AECs ALI induced by LPS and ischemia-reperfusion injury, respectively [65,66]. miR-615-5p is an

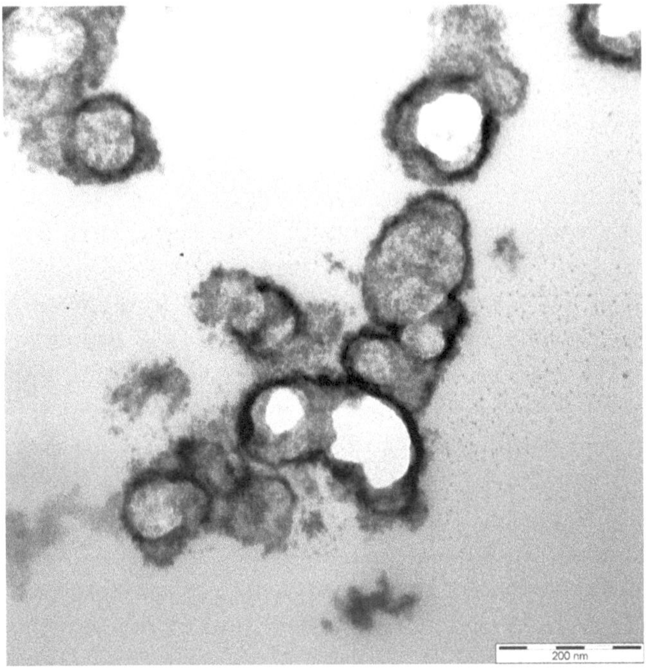

Figure 3.3 Electron microscopic image of cryo-fixed, highly purified exosome preparation from mesenchymal stem cells.

antiangiogenic microRNA targeting IGF2 (insulin-like growth factor 2) and RASSF2 (Ras-associating domain family member 2) that interfered with eNOS (endothelial nitric oxide synthase) signaling which contributes to endothelial leakage [67]. Furthermore, nine miRNA molecules identified by bioinformatic analysis to be complementary to the SARS-Cov-2 viral genome could be candidates to interfere with viral RNA transcription or protein translation essential for viral replication [68,69]. An example of miRNAs identified from EVs of placental derived mesenchymal stem cells is shown Fig. 3.3.

Modulation of the hyperinflammatory immune response through reduction of inflammatory mediators, increased expression ofanti-inflammatory mediators, decreased influx of inflammatory cells (neutrophils, M1 macrophages), increased polarization into M2 macrophages providing high capacity to eliminate tissue debris (Fig. 3.3), and differentiation of a regulatory T cell phenotype could arrest the progression of the lung damage in COVID-19. Consequently, intercellular junctions between alveolar cells and pulmonary capillary endothelial cells could be regenerated along with AEC repair. Degradation of hyaluronic acid accumulating in ARDS lungs and improved mucociliary activity are likely to improve clearance of alveolar fluid and debris. Also, if the nine human miRNA molecules found to be complementary to the SARS-Cov-2 RNA genome can demonstrate any direct or indirect activity to inhibit viral replication, as in the cases of other RNA viruses such as influenza and hepatitis C, MSC-sEVs could interfere with viral replication [71–78].

Investigation of MSC-sEVs in preclinical models of ALI and ARDS have demonstrated significant reduction in inflammatory mediators, inflammatory cell influx, AEC apoptosis, bacterial load, and viral replication, as well as significant improvement in anti-inflammatory mediators, alveolar-capillary barrier permeability, monocyte phagocytosis and alveolar-arterial oxygen gradient [25,28,29,79–81]. Preclinical studies of MSC exosomes as therapy for influenza virus-induced ALI (similar to COVID-19 pulmonary disease) in a clinically relevant swine model demonstrated inhibition of virus-induced apoptosis in AECs, inhibition of influenza virus replication through miRNA transfer, decreased virus shedding, decreased virus replication in the lungs, and decreased production of proinflammatory cytokines [82].

Early clinical studies demonstrating 100% clinical recovery rate from COVID-19 with systemic administration of adipose-derived stromal progenitor cells and placental MSCs, whose mechanism of action is transfer of EVs, also suggest potential efficacy of MSC-sEVs produced by these types of cells. One small cohort study in humans demonstrated direct evidence of a therapeutic benefit of MSC-sEVs in the treatment of COVID-19 with significant improvements in absolute neutrophil count, lymphopenia and acute phase reactants including C-reactive protein, ferritin, and D-dimer. Another phase I study of exosomes bearing CD24 also demonstrated complete recovery in all 30 patients with moderate to severe COVID-19, and recovery within three to five days in 29 of these patients. (https://www.jpost.com/health-science/tel-aviv-hospital-cures-29-of-30-covid-19-patients-in-days-it-says-658024)

8. Source of exosomes

Numerous peer-reviewed preclinical and clinical studies have demonstrated promising therapeutic bioactivity of MSC exosomes for more hundreds of different clinical indications including COVID-19 and acute lung injury [83—86]. The vast majority of these studies have focused on bone marrow-derived MSC (BM-MSC) exosomes, perhaps because they were the first type to be isolated. The remainder of these studies have evaluated adipose tissue- (AT-MSC), umbilical cord- (UC-MSC) and placenta-derived (P-MSC) exosomes to a lesser extent [90]. Comparative studies of these types of MSC exosomes have demonstrated that their cargo and bioactivity differ significantly [91]. It is important to note that the type and state of the producer cells significantly influences the bioactivity of the exosomes they release. Placental MSCs have demonstrated greater immunomodulatory effects and regenerative capacity than bone marrow, adipose and umbilical cord derived MSCs [92,93]. Also, whereas placental MSCs are isolated from perinatal donor tissue usually discarded as medical waste, bone marrow- and adipose-derived MSCs are most commonly isolated from adult donor tissue. Perhaps as a result of changes associated with aging or environmental exposure, BM-MSC and AT-MSC exosomes have demonstrated some tumorigenic and pro-metastatic effects in preclinical studies [94, 95, 96, 97]. As suggested by preclinical studies in which RNAse treatment of MSC exosomes abrogates their bioactivity, this malignant behavior may correlate with the miRNA cargo of these types of MSC exosomes. The miRNA content of BM-MSC exosomes differs significantly from that of P-MSC exosomes [98]. In the case of BM-MSC exosomes, the most abundant microRNA species is miR-1246, which has demonstrated significant oncogenic and metastatic effects. This microRNA species is not present in biologically significant quantities in P-MSC exosomes. Instead, the abundant miRNAs identified in PMSC exosomes participate in important tissue reconstitution, anti-inflammatory pathways and tumor suppression [98]. MSC exosomes hold tremendous potential for the reconstitution of natural tissue barriers, such as the alveolar-capillary barrier that is disrupted in COVID-19, as well as those barriers disrupted in many other clinical conditions. Knowing that the producer cell type and environment strongly influence the character and bioactivity of the exosomes secreted, the development of exosome-based therapeutics for conditions such as COVID-19 and acute lung injury will require optimal producer cell selection, standardized biomanufacturing protocols and rigorous quality management standards. Orthogonal exosome characterization methods including mass spectrometry proteomics and lipidomics, RNA sequencing, nano-flow cytometry and electron microscopy will help to confirm optimal purity and consistency of these products and enable the development and eventual clinical use upon regulatory approval of this next generation of biopharmaceuticals.

The miRNAs of certain mesenchymal stem cell derived exosomes participate in important tissue reconstitution and anti-inflammatory pathways (Fig. 3.3).

9. Mesenchymal stem cell-small extracellular vesicles as investigational new drug

Isolated MSC-sEVs derived from cultures of P-MSCs of fetal origin can be primed with an optimal combination of IL-6, IFN-γ, IL-1β, and Poly (I:C) or overexpressing specific miRNA could be an excellent drug candidate for clinical trials such as for COVID-19. The investigational study participant population could be inclusive of for patients suffering from post COVID-19 disease, because MSC exosomes may be able to arrest the progression of fibrosis in pulmonary disease, and may influence the detrimental progression of autoimmunity [99].

10. Exosome enrichment

The process of exosome enrichment for in-vitro, ex-vivo (whole blood assays), and the application in-vivo requires highly purified and well-defined material. Among different biochemical and molecular methods, electron microscopy performed under most stringent conditions of biological structure preservation is highly valid (Fig. 3.4).

In vivo, exosomes may attenuate inflammation and eventually reconstitute endothelial barrier function in acute respiratory distress syndromes, primarily by reconstituting the balance of pro- (M1) and antiinflammatory macrophages (M2) as schematically demonstrated (Fig. 3.3), In this illustration the preferential release of exosomes targeted by MSC-derived exosomes may constitute a highly relevant amplification mechanism occuring at the site of damage and inflammation. This local amplification mechanism would be followed by resolution of thrombotic elements in small vessels of other organs as well.

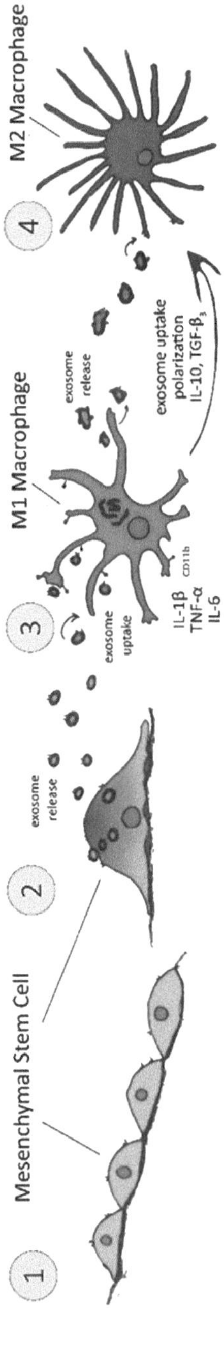

Figure 3.4 COVID-19 lung damage and acute respiratory distress syndromes.

References

[1] Wang C, Liu Z, Chen Z, Huang X, Xu M, He T, Zhang Z. The establishment of reference sequence for SARS-CoV-2 and variation analysis. J Med Virol 2020;92(6):667–74. https://doi.org/10.1002/jmv.25762.

[2] Lukassen S, Chua RL, Trefzer T, et al. SARS-CoV-2 receptor ACE2 and TMPRSS2 are primarily expressed in bronchial transient secretory cells. EMBO J 2020,39(10). e105114. https://doi.org/10.15252/embj.20105114.

[3] Mossel EC, Wang J, Jeffers S, Edeen KE, Wang S, Cosgrove GP, et al. SARS-CoV replicates in primary human alveolar type II cell cultures but not in type I-like cells. Virology 2008;372(1):127–35. https://doi.org/10.1016/j.virol.2007.09.045.

[4] Zhou F, Yu T, Du R, Fan G, Liu Y, Liu Z, et al. Clinical course and risk factors for mortality of adult inpatients with COVID-19 in Wuhan, China: a retrospective cohort study. Lancet (London, England) 2020;395(10229):1054–62. https://doi.org/10.1016/S0140-6736(20)30566-3.

[5] Leiva-Juárez MM, Kolls JK, Evans SE. Lung epithelial cells: therapeutically inducible effectors of antimicrobial defense. Mucosal Immunol 2018;11(1):21–34. https://doi.org/10.1038/mi.2017.71.

[6] Cao X. COVID-19: immunopathology and its implications for therapy. Nat Rev Immunol 2020;20(5):269–70. https://doi.org/10.1038/s41577-020-0308-3.

[7] Ksiazek TG, Erdman D, Goldsmith CS, Zaki SR, Peret T, Emery S, et al. A novel coronavirus associated with severe acute respiratory syndrome. N Engl J Med 2003;348(20):1953–66. https://doi.org/10.1056/NEJMoa030781.

[8] He X, Lau E, Wu P, Deng X, Wang J, Hao X, et al. Temporal dynamics in viral shedding and transmissibility of COVID-19. Nat Med 2020;26(5):672–5. https://doi.org/10.1038/s41591-020-0869-5.

[9] Chu H, Chan JF, Wang Y, Yuen TT, Chai Y, Hou Y, et al. Comparative replication and immune activation profiles of SARS-CoV-2 and SARS-CoV in human lungs: an ex vivo study with implications for the pathogenesis of COVID-19. Clinical infectious diseases : an official publication of the Infectious Diseases Society of America, ciaa410. Advance online publication; 2020. https://doi.org/10.1093/cid/ciaa410.

[10] Chen N, Zhou M, Dong X, Qu J, Gong F, Han Y, et al. Epidemiological and clinical characteristics of 99 cases of 2019 novel coronavirus pneumonia in Wuhan, China: a descriptive study. Lancet (London, England) 2020;395(10223):507–13. https://doi.org/10.1016/S0140-6736(20)30211-7.

[11] National Research Project for SARS, Beijing Group. The involvement of natural killer cells in the pathogenesis of severe acute respiratory syndrome. Am J Clin Pathol 2004;121(4):507–11. https://doi.org/10.1309/WPK7-Y2XK-NF4C-BF3R.

[12] Herold T, Jurinovic V, Arnreich C, Hellmuth J, Bergwelt-Baildon M, Klein M, Weinberger T. Level of IL-6 predicts respiratory failure in hospitalized symptomatic COVID-19 patients. 2020. https://doi.org/10.1101/2020.04.01.20047381.

[13] Conti P, Ronconi G, Caraffa A, et al. Induction of pro-inflammatory cytokines (IL-1 and IL-6) and lung inflammation by Coronavirus-19 (COVI-19 or SARS-CoV-2): anti-inflammatory strategies. J Biol Regul Homeost Agents 2020;34(2):327–31. https://doi.org/10.23812/CONTI-E.

[14] Lang Z, Zhang L, Zhang S, Meng X, Li J, Song C, et al. Pathological study on severe acute respiratory syndrome. Chin Med J 2003;116(7):976–80.

[15] Li H, Meng X, Liang X, Gao Y, Cai S. Administration of microparticles from blood of the lipopolysaccharide-treated rats serves to induce pathologic changes of acute respiratory distress syndrome. Exp Biol Med 2015;240(12):1735–41. https://doi.org/10.1177/1535370215591830.

[16] Qian Z, Travanty EA, Oko L, Edeen K, Berglund A, Wang J, et al. Innate immune response of human alveolar type II cells infected with severe acute respiratory syndrome-coronavirus. Am J Respir Cell Mol Biol 2013;48(6):742–8. https://doi.org/10.1165/rcmb.2012-0339OC.

[17] Abbas AK, Lichtman AH, Pillai S. Cellular and molecular immunology. 9th ed. Philadelphia: Saunders Elsevier; 2017.

[18] Yoshikawa T, Hill T, Li K, Peters CJ, Tseng CT. Severe acute respiratory syndrome (SARS) coronavirus-induced lung epithelial cytokines exacerbate SARS pathogenesis by modulating intrinsic functions of monocyte-derived macrophages and dendritic cells. J Virol 2009;83(7):3039–48. https://doi.org/10.1128/JVI.01792-08.

[19] Gu J, Korteweg C. Pathology and pathogenesis of severe acute respiratory syndrome. Am J Pathol 2007;170(4):1136–47. https://doi.org/10.2353/ajpath.2007.061088.

[20] Fang X, Neyrinck AP, Matthay MA, Lee JW. Allogeneic human mesenchymal stem cells restore epithelial protein permeability in cultured human alveolar type II cells by secretion of angiopoietin-1. J Biol Chem 2010;285(34):26211–22. https://doi.org/10.1074/jbc.M110.119917.

[21] van de Veerdonk F, Netea MG, van Deuren M, van der Meer JW, de Mast Q, Bruggemann RJ, van der Hoeven H. Kinins and cytokines in COVID-19: a comprehensive pathophysiological approach. Preprints 2020:2020040023. https://doi.org/10.20944/preprints202004.0023.v1.

[22] Fujio K, Okamura T, Sumitomo S, Yamamoto K. Therapeutic potential of regulatory cytokines that target B cells. Int Immunol 2016;28(4):189–95. https://doi.org/10.1093/intimm/dxv069.

[23] Chuang Y, Hung ME, Cangelose BK, Leonard JN. Regulation of the IL-10-driven macrophage phenotype under incoherent stimuli. Innate Immun 2016;22(8):647–57. https://doi.org/10.1177/1753425916668243.

[24] Monsel A, Zhu YG, Gudapati V, Lim H, Lee JW. Mesenchymal stem cell derived secretome and extracellular vesicles for acute lung injury and other inflammatory lung diseases. Expet Opin Biol Ther 2016;16(7):859–71. https://doi.org/10.1517/14712598.2016.1170804.

[25] Morrison TJ, Jackson MV, Cunningham EK, Kissenpfennig A, McAuley DF, O'Kane CM, Krasnodembskaya AD. Mesenchymal stromal cells modulate macrophages in clinically relevant lung injury models by extracellular vesicle mitochondrial transfer. Am J Respir Crit Care Med 2017;196(10):1275–86. https://doi.org/10.1164/rccm.201701-0170OC.

[26] Komai T, Okamura T, Inoue M, Yamamoto K, Fujio K. Reevaluation of pluripotent cytokine TGF-β3 in immunity. Int J Mol Sci 2018;19(8):2261. https://doi.org/10.3390/ijms19082261.

[27] Goolaerts A, Pellan-Randrianarison N, Larghero J, Vanneaux V, Uzunhan Y, Gille T, et al. Conditioned media from mesenchymal stromal cells restore sodium transport and preserve epithelial permeability in an in vitro model of acute alveolar injury. Am J Physiol Lung Cell Mol Physiol 2014;306(11):L975–85. https://doi.org/10.1152/ajplung.00242.2013.

[28] Hu S, Park J, Liu A, Lee J, Zhang X, Hao Q, Lee JW. Mesenchymal stem cell microvesicles restore protein permeability across primary cultures of injured human lung microvascular endothelial cells. Stem Cells Transl Med 2018;7(8):615–24. https://doi.org/10.1002/sctm.17-0278.

[29] Tang XD, Shi L, Monsel A, Li XY, Zhu HL, Zhu YG, Qu JM. Mesenchymal stem cell microvesicles attenuate acute lung injury in mice partly mediated by Ang-1 mRNA. Stem Cells (Dayton, Ohio) 2017;35(7):1849–59. https://doi.org/10.1002/stem.2619.

[30] Lee JW, Krasnodembskaya A, McKenna DH, Song Y, Abbott J, Matthay MA. Therapeutic effects of human mesenchymal stem cells in ex vivo human lungs injured with live bacteria. Am J Respir Crit Care Med 2013;187(7):751–60. https://doi.org/10.1164/rccm.201206-0990OC.

[31] Mei SH, McCarter SD, Deng Y, Parker CH, Liles WC, Stewart DJ. Prevention of LPS-induced acute lung injury in mice by mesenchymal stem cells overexpressing angiopoietin 1. PLoS Med 2007;4(9):e269. https://doi.org/10.1371/journal.pmed.0040269.

[32] Mei SH, Haitsma JJ, Dos Santos CC, Deng Y, Lai PF, Slutsky AS, et al. Mesenchymal stem cells reduce inflammation while enhancing bacterial clearance and improving survival in sepsis. Am J Respir Crit Care Med 2010;182(8):1047–57. https://doi.org/10.1164/rccm.201001-0010OC.

[33] Nemeth K, Mayer B, Mezey E. Modulation of bone marrow stromal cell functions in infectious diseases by toll-like receptor ligands. J Mol Med (Berlin, Germany) 2010;88(1):5–10. https://doi.org/10.1007/s00109-009-0523-7.

[34] Krasnodembskaya A, Song Y, Fang X, Gupta N, Serikov V, Lee JW, Matthay MA. Antibacterial effect of human mesenchymal stem cells is mediated in part from secretion of the antimicrobial peptide LL-37. Stem Cells (Dayton, Ohio) 2010;28(12):2229–38. https://doi.org/10.1002/stem.544.

[35] Lee JW, Fang X, Gupta N, Serikov V, Matthay MA. Allogeneic human mesenchymal stem cells for treatment of *E. coli* endotoxin-induced acute lung injury in the ex vivo perfused human lung. Proc Natl Acad Sci U S A 2009;106(38):16357–62. https://doi.org/10.1073/pnas.0907996106.

[36] Prockop DJ, Oh JY. Mesenchymal stem/stromal cells (MSCs): role as guardians of inflammation. Mol Ther 2012;20(1):14–20. https://doi.org/10.1038/mt.2011.211. The Journal of the American Society of Gene Therapy.

[37] Leng Z, Zhu R, Hou W, Feng Y, Yang Y, Han Q, et al. Transplantation of ACE2-mesenchymal stem cells improves the outcome of patients with COVID-19 pneumonia. Aging Dis 2020;11(2):216–28. https://doi.org/10.14336/AD.2020.0228.

[38] https://www.jpost.com/health-science/israeli-cell-therapy-to-treat-acute-covid-19-patients-prevent-ards-621016.

[39] Witwer K, Van Balkom B, Bruno S, Choo A, Dominici M, Gimona M, et al. Defining mesenchymal stromal cell (MSC)-derived small extracellular vesicles for therapeutic applications. J Extracell Vesicles 2019;8:1609206. https://doi.org/10.1080/20013078.2019.1609206.

[40] Lener T, Gimona M, Aigner L, Börger V, Buzas E, Camussi G, et al. Applying extracellular vesicles based therapeutics in clinical trials – an ISEV position paper. J Extracell Vesicles 2015;4:30087. https://doi.org/10.3402/jev.v4.30087.

[41] Börger V, Bremer M, Ferrer-Tur R, Gockeln L, Stambouli O, Becic A, Giebel B. Mesenchymal stem/stromal cell-derived extracellular vesicles and their potential as novel immunomodulatory therapeutic agents. Int J Mol Sci 2017;18(7):1450. https://doi.org/10.3390/ijms18071450.

[42] Lin KC, Yip HK, Shao PL, Wu SC, Chen KH, Chen YT, et al. Combination of adipose-derived mesenchymal stem cells (ADMSC) and ADMSC-derived exosomes for protecting kidney from acute ischemia-reperfusion injury. Int J Cardiol 2016;216:173–85. https://doi.org/10.1016/j.ijcard.2016.04.0611.

[43] Nong K, Wang W, Niu X, Hu B, Ma C, Bai Y, et al. Hepatoprotective effect of exosomes from human-induced pluripotent stem cell-derived mesenchymal stromal cells against hepatic ischemia-reperfusion injury in rats. Cytotherapy 2016;18(12):1548–59. https://doi.org/10.1016/j.jcyt.2016.08.002.

[44] Zou X, Zhang G, Cheng Z, Yin D, Du T, Ju G, et al. Microvesicles derived from human Wharton's Jelly mesenchymal stromal cells ameliorate renal ischemia-reperfusion injury in rats by suppressing CX3CL1. Stem Cell Res Ther 2014;5(2):40. https://doi.org/10.1186/scrt428.
[45] Chen L, Xiang B, Wang X, et al. Exosomes derived from human menstrual blood-derived stem cells alleviate fulminant hepatic failure. Stem Cell Res Ther 2017;8:9. https://doi.org/10.1186/s13287-016-0453-6.
[46] Budoni M, Fierabracci A, Luciano R, Petrini S, Di Ciommo V, Muraca M. The immunosuppressive effect of mesenchymal stromal cells on B lymphocytes is mediated by membrane vesicles. Cell Transpl 2013;22(2):369–79. https://doi.org/10.3727/096368911X582769.
[47] Blazquez R, Sanchez-Margallo FM, de la Rosa O, Dalemans W, Alvarez V, Tarazona R, Casado JG. Immunomodulatory potential of human adipose mesenchymal stem cells derived exosomes on in vitro stimulated T cells. Front Immunol 2014;5:556. https://doi.org/10.3389/fimmu.2014.00556.
[48] Mokarizadeh A, Delirezh N, Morshedi A, Mosayebi G, Farshid AA, Mardani K. Microvesicles derived from mesenchymal stem cells: potent organelles for induction of tolerogenic signaling. Immunol Lett 2012;147(1–2):47–54. https://doi.org/10.1016/j.imlet.2012.06.001.
[49] Del Fattore A, Luciano R, Pascucci L, Goffredo BM, Giorda E, Scapaticci M, et al. Immunoregulatory effects of mesenchymal stem cell-derived extracellular vesicles on T lymphocytes. Cell Transpl 2015;24(12):2615–27. https://doi.org/10.3727/096368915X687543.
[50] Favaro E, Carpanetto A, Lamorte S, Fusco A, Caorsi C, Deregibus MC, et al. Human mesenchymal stem cell-derived microvesicles modulate T cell response to islet antigen glutamic acid decarboxylase in patients with type 1 diabetes. Diabetologia 2014;57(8):1664–73. https://doi.org/10.1007/s00125-014-3262-4.
[51] Conforti A, Scarsella M, Starc N, Giorda E, Biagini S, Proia A, et al. Microvescicles derived from mesenchymal stromal cells are not as effective as their cellular counterpart in the ability to modulate immune responses in vitro. Stem Cells Dev 2014;23(21):2591–9. https://doi.org/10.1089/scd.2014.0091.
[52] Lo Sicco C, Reverberi D, Balbi C, Ulivi V, Principi E, Pascucci L, et al. Mesenchymal stem cell-derived extracellular vesicles as mediators of anti-inflammatory effects: endorsement of macrophage polarization. Stem Cells Transl Med 2017;6(3):1018–28. https://doi.org/10.1002/sctm.16-0363.
[53] Ti D, Hao H, Tong C, Liu J, Dong L, Zheng J, et al. LPS-preconditioned mesenchymal stromal cells modify macrophage polarization for resolution of chronic inflammation via exosome-shuttled let-7b. J Transl Med 2015;13:308. https://doi.org/10.1186/s12967-015-0642-6.
[54] Guan YT, Xie Y, Li DS, Zhu YY, Zhang XL, Feng YL, et al. Comparison of biological characteristics of mesenchymal stem cells derived from the human umbilical cord and decidua parietalis. Mol Med Rep 2019;20(1):633–9. https://doi.org/10.3892/mmr.2019.10286.
[55] Shao M, Xu Q, Wu Z, et al. Exosomes derived from human umbilical cord mesenchymal stem cells ameliorate IL-6-induced acute liver injury through miR-455-3p. Stem Cell Res Ther 2020;11:37. https://doi.org/10.1186/s13287-020-1550-0.
[56] Su YL, Wang X, Mann M, Adamus TP, Wang D, Moreira DF, et al. Myeloid cell-targeted miR-146a mimic inhibits NF-κB-driven inflammation and leukemia progression in vivo. Blood 2020;135(3):167–80. https://doi.org/10.1182/blood.2019002045.

[57] Shao Q, Liu F, Zhan Y, Nie C, Zhu W, Qian K. Upregulation of miR-146a contributes to the suppression of inflammatory responses in LPS-induced acute lung injury. Exp Lung Res 2013;39(7):275–82. https://doi.org/10.3109/01902148.2013.808285.
[58] Zeng Z, Gong H, Li Y, Jie K, Ding C, Shao Q, et al. Upregulation of miR-146a contributes to the suppression of inflammatory responses in LPS-induced acute lung injury. Exp Lung Res 2013;39(7):275–82. https://doi.org/10.3109/01902148.2013.808285.
[59] Bhaumik D, Scott GK, Schokrpur S, Patil CK, Orjalo AV, Rodier F, et al. MicroRNAs miR-146a/b negatively modulate the senescence-associated inflammatory mediators IL-6 and IL-8. Aging 2009;1(4):402–11. https://doi.org/10.18632/aging.100042.
[60] Song Y, Dou H, Li X, Zhao X, Li Y, Liu D, et al. Exosomal miR-146a contributes to the enhanced therapeutic efficacy of interleukin-1β-primed mesenchymal stem cells against sepsis. Stem Cells (Dayton, Ohio) 2017;35(5):1208–21. https://doi.org/10.1002/stem.2564.
[61] Zhuang G, et al. A novel regulator of macrophage activation: miR-223 in obesity-associated adipose tissue inflammation. Circulation 2012;125:2892–903.
[62] Wu J, Niu P, Zhao Y, Cheng Y, Chen W, Lin L, et al. Impact of miR-223-3p and miR-2909 on inflammatory factors IL-6, IL-1ß, and TNF-α, and the TLR4/TLR2/NF-κB/STAT3 signaling pathway induced by lipopolysaccharide in human adipose stem cells. PLoS One 2019;14(2):e0212063. https://doi.org/10.1371/journal.pone.0212063.
[63] Squadrito ML, Pucci F, Magri L, Moi D, Gilfillan GD, Ranghetti A, et al. miR-511-3p modulates genetic programs of tumor-associated macrophages. Cell Rep 2012;1(2):141–54. https://doi.org/10.1016/j.celrep.2011.12.005.
[64] Chen WX, Zhou J, Zhou SS, Zhang YD, Ji TY, Zhang XL, et al. Microvesicles derived from human Wharton's jelly mesenchymal stem cells enhance autophagy and ameliorate acute lung injury via delivery of miR-100. Stem Cell Res Ther 2020;11(1):113. https://doi.org/10.1186/s13287-020-01617-7.
[65] Yi X, Wei X, Lv H, An Y, Li L, Lu P, et al. Exosomes derived from microRNA-30b-3p-overexpressing mesenchymal stem cells protect against lipopolysaccharide-induced acute lung injury by inhibiting SAA3. Exp Cell Res 2019;383(2):111454. https://doi.org/10.1016/j.yexcr.2019.05.035.
[66] Li JW, Wei L, Han Z, Chen Z. Mesenchymal stromal cells-derived exosomes alleviate ischemia/reperfusion injury in mouse lung by transporting anti-apoptotic miR-21-5p. Eur J Pharmacol 2019;852:68–76. https://doi.org/10.1016/j.ejphar.2019.01.022.
[67] Icli B, Wu W, Ozdemir D, Li H, Cheng HS, Haemmig S, et al. MicroRNA-615-5p regulates angiogenesis and tissue repair by targeting AKT/eNOS (protein kinase B/endothelial nitric oxide synthase) signaling in endothelial cells. Arterioscler Thromb Vasc Biol 2019;39(7):1458–74. https://doi.org/10.1161/ATVBAHA.119.312726.
[68] Sardar R, Satish D, Birla S, Gupta D. Comparative analyses of SAR-CoV2 genomes from different geographical locations and other coronavirus family genomes reveals unique features potentially consequential to host-virus interaction and pathogenesis. 2020. 10.1101/2020.03.21.001586.
[69] Demirci M, Adan A. Computational analysis of microRNA-mediated interactions in SARS-CoV-2 infection. bioRxiv 2020;03(15):992438. https://doi.org/10.1101/2020.03.15.992438.
[70] Abraham A, Krasnodembskaya A. Mesenchymal stem cell-derived extracellular vesicles for the treatment of acute respiratory distress syndrome. Stem Cells Transl Med 2020;9(1):28–38. https://doi.org/10.1002/sctm.19-02055.

[71] Song L, Liu H, Gao S, Jiang W, Huang W. Cellular microRNAs inhibit replication of the H1N1 influenza A virus in infected cells. J Virol 2010;84(17):8849–60. https://doi.org/10.1128/JVI.00456-10.

[72] Ingle H, Kumar S, Raut AA, Mishra A, Kulkarni DD, Kameyama T, et al. The microRNA miR-485 targets host and influenza virus transcripts to regulate antiviral immunity and restrict viral replication. Sci Signal 2015;8(406):ra126. https://doi.org/10.1126/scisignal.aab3183.

[73] Khongnomnan K, Makkoch J, Poomipak W, Poovorawan Y, Payungporn S. Human miR-3145 inhibits influenza A viruses replication by targeting and silencing viral PB1 gene. Exp Biol Med 2015;240(12):1630–9. https://doi.org/10.1177/1535370215589051.

[74] Ma YJ, Yang J, Fan XL, Zhao HB, Hu W, Li ZP, et al. Cellular microRNA let-7c inhibits M1 protein expression of the H1N1 influenza A virus in infected human lung epithelial cells. J Cell Mol Med 2012;16(10):2539–46. https://doi.org/10.1111/j.1582-4934.2012.01572.x.

[75] Sun E, He J, Zhuang X. Dissecting the role of COPI complexes in influenza virus infection. J Virol 2013;87(5):2673–85. https://doi.org/10.1128/JVI.02277-12.

[76] König R, Stertz S, Zhou Y, Inoue A, Hoffmann HH, Bhattacharyya S, et al. Human host factors required for influenza virus replication. Nature 2010;463(7282):813–7. https://doi.org/10.1038/nature08699.

[77] Waring BM, Sjaastad LE, Fiege JK, Fay EJ, Reyes I, Moriarity B, Langlois RA. MicroRNA-Based attenuation of influenza virus across susceptible hosts. J Virol 2018;92(2). https://doi.org/10.1128/JVI.01741-17. e01741–17.

[78] Qian X, Xu C, Fang S, Zhao P, Wang Y, Liu H, et al. Exosomal MicroRNAs derived from umbilical mesenchymal stem cells inhibit hepatitis C virus infection. Stem Cells Transl Med 2016;5(9):1190–203. https://doi.org/10.5966/sctm.2015-0348.

[79] Zhu YG, Feng XM, Abbott J, Fang XH, Hao Q, Monsel A, et al. Human mesenchymal stem cell microvesicles for treatment of *Escherichia coli* endotoxin-induced acute lung injury in mice. Stem Cells (Dayton, Ohio) 2014;32(1):116–25. https://doi.org/10.1002/stem.1504.

[80] Gennai S, Monsel A, Hao Q, Park J, Matthay MA, Lee JW. Microvesicles derived from human mesenchymal stem cells restore alveolar fluid clearance in human lungs rejected for transplantation. Am J Transplant 2015;15(9):2404–12. https://doi.org/10.1111/ajt.13271. Official Journal of the American Society of Transplantation and the American Society of Transplant Surgeons.

[81] Park J, Kim S, Lim H, Liu A, Hu S, Lee J, et al. Therapeutic effects of human mesenchymal stem cell microvesicles in an ex vivo perfused human lung injured with severe *E. coli* pneumonia. Thorax 2019;74(1):43–50. https://doi.org/10.1136/thoraxjnl-2018-211576.

[82] Khatri M, Richardson LA, Meulia T. Mesenchymal stem cell-derived extracellular vesicles attenuate influenza virus-induced acute lung injury in a pig model. Stem Cell Res Ther 2018;9(1):17. https://doi.org/10.1186/s13287-018-0774-8.

[83] Li CY, Wu XY, Tong JB, Yang XX, Zhao JL, Zheng QF, et al. Comparative analysis of human mesenchymal stem cells from bone marrow and adipose tissue under xeno-free conditions for cell therapy. Stem Cell Res Ther 2015;6(1):55. https://doi.org/10.1186/s13287-015-0066-5.

[84] Huang Y, Liu W, He B, Wang L, Zhang F, Shua H, Sun L. Exosomes derived from bone marrow mesenchymal stem cells promote osteosarcoma development by activating oncogenic autophagy. J Bone Oncol 2020;(21). https://doi.org/10.1016/j.jbo.2020.100280.

[85] Zhang X, Xu Y, Wang J, Zhao S, Li J, Huang X, et al. miR-221-3p delivered by BMMSC-derived microvesicles promotes the development of acute myelocytic leukemia. Front Bioeng Biotechnol 2020;8:81. https://doi.org/10.3389/fbioe.2020.00081.

[86] Wang Q, Li Z, Sun L, Chen B, Zhao Y, Shen B, et al. Platelets enhance the ability of bone-marrow mesenchymal stem cells to promote cancer metastasis. OncoTargets Ther 2018;11:8251–63. https://doi.org/10.2147/OTT.S181673.

[87] Huang Q, Yang Y, Luo C, Wen Y, Liu R, Li S, et al. An efficient protocol to generate placental chorionic plate-derived mesenchymal stem cells with superior proliferative and immunomodulatory properties. Stem Cell Res Ther 2019;10. 10.1186/s13287-019-1405-8.

[88] Monsel A, Zhu YG, Gennai S, Hao Q, Hu S, Rouby JJ, et al. Therapeutic effects of human mesenchymal stem cell-derived microvesicles in severe pneumonia in mice. Am J Respir Crit Care Med 2015;192(3):324–36. https://doi.org/10.1164/rccm.201410-1765OC.

[89] Varkouhi AK, Jerkic M, Ormesher L, Gagnon S, Goyal S, Rabani R, et al. Extracellular vesicles from interferon-γ-primed human umbilical cord mesenchymal stromal cells reduce *Escherichia coli*-induced acute lung injury in rats. Anesthesiology 2019; 130(5):778–90. https://doi.org/10.1097/ALN.0000000000002655.

[90] Stem cells and development 2020. https://doi.org/10.1089/scd.2020.0080.

[91] Hicoik K. Stem Cells Regen Med 2020;4(1):1–4.

[92] International immunopharmacology 2012. https://doi.org/10.1016/j.intimp.2012.03.024.

[93] Stem cell research & therapy 2019. https://doi.org/10.1186/s13287-019-1405-8.

[94] Jin. Int J Mol Sci 2013.

[95] Huang. Journal of Bone Oncology 2020.

[96] Zhang. Front. Bioeng. Biotechnol 2020.

[97] Wang. Onco Targets Ther 2018.

[98] Ferguson. Scientific reports 2018. https://doi.org/10.1038/s41598-018-19581-x.

[99] Zuo Y. Prothrombotic autoantibodies in serum from patients hospitalized with COVID-19. Science Translational Medicine 2020;12(570). https://doi.org/10.1126/scitranslmed.abd3876.

CHAPTER FOUR

The retinal pigment epithelium: at the forefront of the blood-retinal barrier in physiology and disease

Richárd Nagymihály[1,a] Yaroslav Nemesh[2,a] Taras Ardan[2], Jan Motlik[2], Jon Roger Eidet[1], Morten C. Moe[1,3] Linda Hildegard Bergersen[4], Lyubomyr Lytvynchuk[5], Goran Petrovski[1,3]

[1]Center for Eye Research, Department of Ophthalmology, Oslo University Hospital, Oslo, Norway
[2]Laboratory of Cell Regeneration and Plasticity, Institute of Animal Physiology and Genetics, Czech Academy of Science, Libechov, Czech Republic
[3]Institute of Clinical Medicine, Faculty of Medicine, University of Oslo, Oslo, Norway
[4]Institute for Oral Biology, Faculty of Dentistry, University of Oslo, Oslo, Norway
[5]Department of Ophthalmology, Justus-Liebig-University Giessen, Eye Clinic, University Hospital Giessen and Marburg GmbH, Giessen, Germany

1. Introduction

The vertebrate retina is a multilayered structure with a large diversity of composing cells that form morphologically and functionally distinct circuits. These circuits work in parallel, and in coherence, to produce a complex visual output [1]. The retina transforms light into chemical signals that are further conveyed to the brain. This process requires an ability to sense light stimulus and transmit signals from cell to cell [2].

Six major cell types form the various layers of the human retina: the outer nuclear layer (ONL) consisting of photoreceptors (rods and cones); the inner nuclear layer (INL) formed mainly of bipolar and amacrine cells; retinal ganglion cells make up the ganglion cell layer (GCL; innermost layer which is farthest from the photoreceptors); horizontal and Müller cells. Ganglion cell axon tracts travel toward the back of the eye and form the optic nerve [2].

The blood-retinal barrier (BRB) plays an important role in maintaining retinal integrity. Similar to the blood-brain barrier (BBB), it protects against toxic chemicals and macromolecules in the capillary blood flow by forming

[a] These authors are shared first authors.

Tissue Barriers in Disease, Injury and Regeneration
ISBN: 978-0-12-818561-2
https://doi.org/10.1016/B978-0-12-818561-2.00003-5

tight control of molecule delivery to the tissue. The BRB consists of two parts: inner and outer layer (Fig. 4.1) [3].

Endothelial cells of the retinal capillaries with their tight junctions (TJs) between and the basal lamina tightly adhere to form the inner BRB (iBRB) [4]. In addition, Müller cell outgrowths, astrocytes, and pericytes surround the endothelial cells and create the whole structure of the iBRB [5,6].

The layer of retinal pigment epithelium (RPE) is located underneath the iBRB, tightly adherent to the Bruch's membrane and interconnected by TJs, together forming the outer BRB (oBRB). The Bruch's membrane is the structural barrier between the neural retina and the choroidal capillaries, which are fenestrated and supply the retina with nutrients [7].

The objective of this chapter is to describe the cellular, molecular, and structural landmarks of the BRB, with particular focus on the RPE at the forefront of this barrier in physiology and disease.

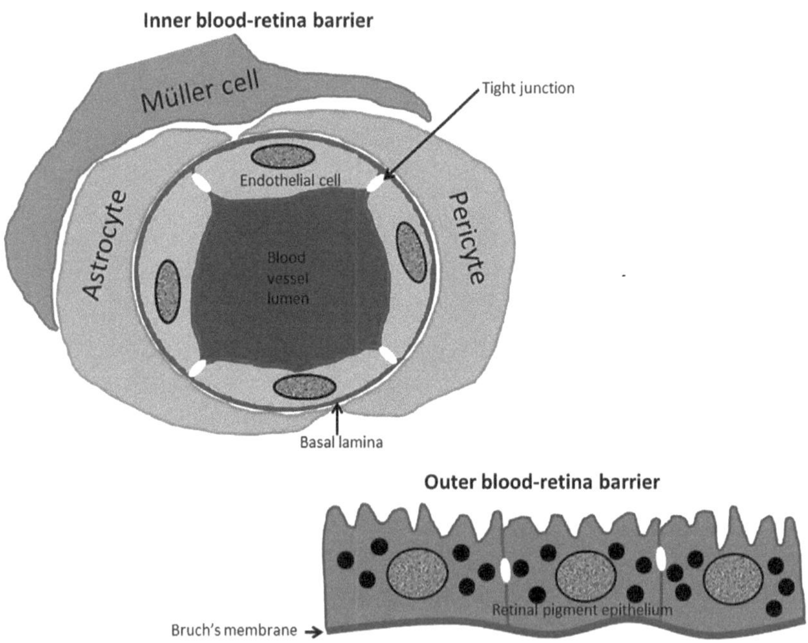

Figure 4.1 Schematic of the inner and outer blood retina barrier constituent cells and structures.

2. Inner blood-retinal barrier

The retinal blood vessels are distributed in two networks: (1) the nerve fibers with the layers of ganglion cells form the superficial (inner) network and (2) the space between the inner nuclear- and the outer plexiform-layers (OPLs) form the deep (outer) network [8].

The retinal nerve fiber layer (RNFL) of the human eye contains additional low vesicle-containing capillaries without fenestrations [8,9]. The vesicles of the endotheliocytes facilitate receptor-mediated or ATP-requiring transcytosis and endocytosis [10,11]. The transcytosis in the endothelial cells occurs by caveolae and is maintained in two possible ways: (1) separate cytoplasm-free caveolae maintain the cargo transport through the cells and (2) the system of fused caveolae form a vesico-tubular system that connects the opposite sides of the cells [12]. Similarly, the endotheliocytes of Schlemm's canal in the anterior eye segment contain a similar vesico-tubular system [13]. Notably, the limited vesicle quantity in the endotheliocytes of the retinal capillaries can facilitate the iBRB continuity and integrity [14].

The endotheliocytes in the retina are sensitive to different damaging factors, such as reactive oxygen species (ROS) that cause increase of cell permeability and pinocytosis. As a result, enhanced generation of vesicles can lead to dramatic alterations with subsequent pathologies in the iBRB [15–17].

2.1 Endothelial cells in the eye vessel walls

TJs between retinal capillary endothelial cells serve as a barrier and prevent the infiltration of molecules across the paracellular space and restrict motion of different lipids and proteins to localize them into the sides of the cells, thus providing plasma membrane polarization into apical and basolateral sides [18,19]. Moreover, TJs take part in the signal transduction, regulation of transcription, proliferation, and differentiation of cells [20,21].

The composition of the TJs varies between two types of proteins—transmembrane and cytoplasmic-scaffold proteins. Transmembrane proteins include claudin and occludin that can control the flow of vascular liquid [22,23], as well as glycoproteins of the cadherins, which provide adhesion of neighboring cells [24,25]. The cytoplasmic proteins in the TJs include the zonula occludens (ZO) -1, -2, -3 (ZO-1, -2, -3), symplekin, 7H6, and cingulin – all playing a role as TJ organizers inside the cell

[23,26,27]. It is important to mention that ZO-1 is the key protein of the TJs which provides the barrier integrity in the retina (Fig. 4.2) [28].

The ciliary body of the eye with its rich blood vessel network is also involved in the barrier function played by the endothelial cells found within (Fig. 4.3). In addition, the vessel walls of larger and smaller vessels and capillaries in the retina contain a barrier built from different types of collagens. Collagen IV expression in these tissues and the sclera is shown here, forming the outer most barrier of the eye (Fig. 4.3). Collagen types I, III, IV, V, and VI have been found in large vessels, types I, IV, and V plus small amounts of III and VI in small vessels, and types I, IV, and V in capillaries. Hyalinized vessel walls have been found to consist mainly of types I, IV, and VI collagen Furthermore, collagen type II has been localized in the vitreous cortex, but not the internal limiting membrane (ILM), while collagen type IV and VI

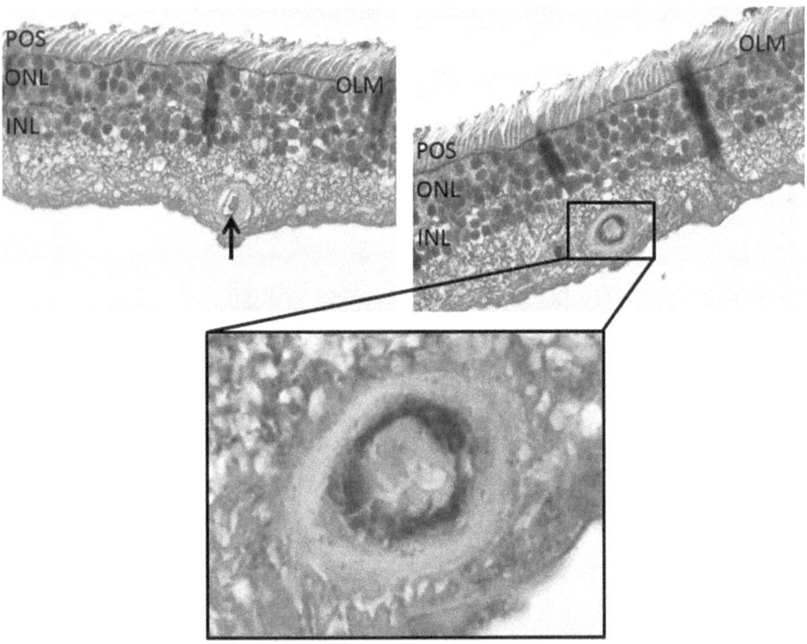

Figure 4.2 Zonula occludens-1 expression in the retinal layers: Cross-section of a human neuroretina stained against Zonula occludens-1 with DAB (brown stain [black in print version]). The selected neuroretinal cell layers seen histologically here consist of: photoreceptor outer segments (POS); outer nuclear layer (ONL); inner nuclear layer (INL). The arrow and the square, together with the enlarged insert show retinal vessel with endothelial cells containing ZO-1 positive tight junctions. Magnification shown in the upper two images: ×63.

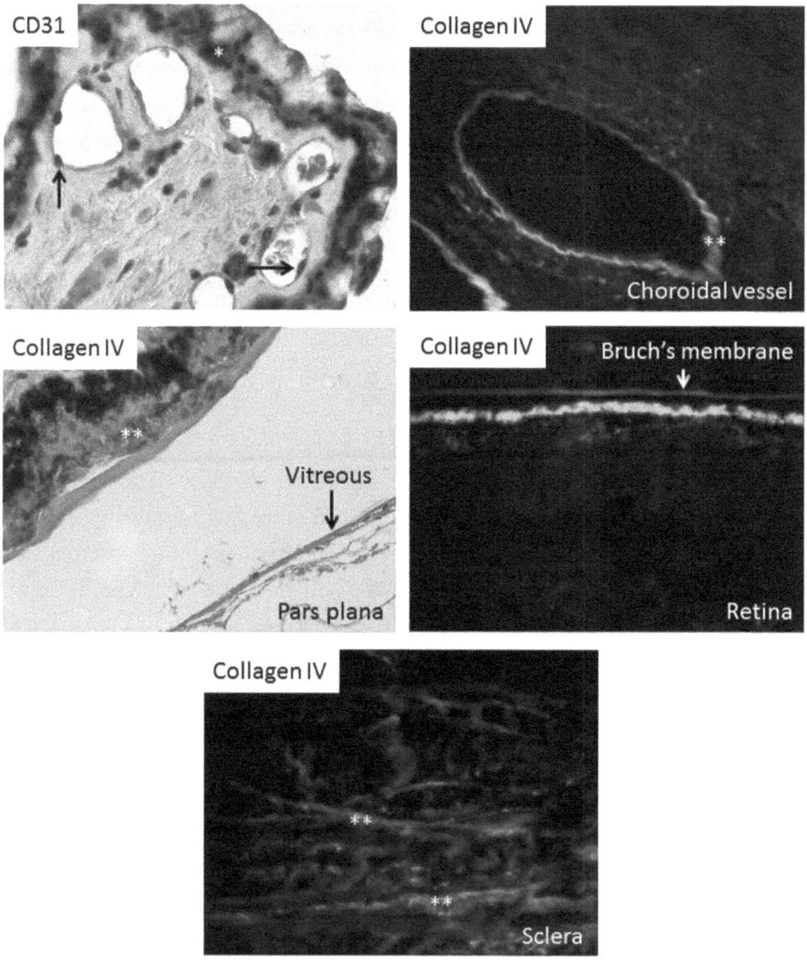

Figure 4.3 Expression of CD31 and collagen IV in the human eye: CD31 expression is shown in the vascular endothelium (arrow) of the pars plicata part of the ciliary body. *Pigmented epithelium of the ciliary body. Collagen IV expression is shown in a large vessel of the choroid (**green fluorescent labeling [gray in print version]), as well as the nonpigmented epithelium of the pars plana region and the vitreous (brown DAB labeling [black in print version]). In addition, the Bruch's membrane of the retina and underneath the highly autofluorescent RPE melanin pigments, as well as the sclera stain positive for collagen IV (**green fluorescent labeling [gray in print version]).

could be found in the ILM and retinal blood vessels, but not the vitreous cortex. Collagen type V could be found in retinal blood vessels and vitreous cortex [29–31].

2.2 Pericytes

The pericytes maintain support of the retinal capillary endotheliocytes by carrying out phagocytosis, extracellular matrix (ECM) secretion, and vessel tone regulation. In addition, pericytes can inhibit endotheliocyte proliferation and participate in angiogenesis [32]. It is remarkable that the number of pericytes in the vessels of the retina is equal to the number of endotheliocytes, and they cover a larger surface area of the capillaries in the retina than in the brain [33–36].

The region of pericytes-capillary endotheliocytes connection is thin enough to permit them to exchange information [33,37]. Pericytes contain a great amount of myosin, tropomyosin, and actin, which allows them to take part in local blood flow regulation by making contractions [32,35,38–41]. There are several factors which make pericytes contract: endothelin-1, angiotensin II, ATP, and hyperoxia [39,42]. The opposite effect is observed under the action of CO_2, NO, and adenosine [43–45].

Pericytes contribute to the iBRB integrity by keeping an appropriate physiological quantity of occludin and ZO-1 under normal conditions and supply occludin during hypoxia [46].

Various pathological conditions can influence iBRB integrity; for instance, diabetes causes partial pericyte degradation with subsequent occludin reduction, as well as TJ breaks and increased permeability of the retinal capillaries [47–51]. Senescence is another reason for a connection loss between pericytes and endotheliocytes, which leads to worsened metabolic products' exchange, which is important for normal neuronal function [52].

2.3 Müller cells

Müller cells, besides the pericytes, are another cell type covering the retinal vessels, which form and maintain the BRB. These cells play a role in maintaining the physiological concentrations of ions, particularly potassium, as well as secretion of signal proteins and regulation of the extracellular pH [53]. Moreover, Müller cells play an important role in the maintenance of neuronal activity, as well as take part in the nutrient supply and metabolite removal to facilitate normal function of the iBRB [6,54,55].

Vascular endothelial growth factor (VEGF), originally known as vascular permeability factor (VPF), is a signal protein of blood vessel formation and permeability increase, which is produced by the Müller cells, as well as astrocytes, vascular endothelium, ganglion cells, and RPEs [56]. To keep balanced angiogenesis, Müller cells also produce pigment epithelium-derived growth factor (PEDF), which is a functional antagonist of the VEGF and acts to reduce vascular permeability [57]. During inflammation and hypoxia, Müller cells secrete an insufficient amount of PEDF that causes an obvious increase in the VEGF and vascular permeability, which result in disruption of the BRB integrity [58–61]. Such disruption is often present in pathological conditions associated with diabetes [60].

The Müller cells can also produce matrix metalloproteinases (MMPs), which can cleave the occludins and cause TJ regression [62,63]. Absence or physical removal of the RPE can, in addition, cause destructive changes in the Müller cells [64], while a local impairment of the Müller cells can result in degradation of retina locally or even more diffusely [65–67]. Müller cells are otherwise robust cell type which can remain alive for a long period time without much oxygen supply [68–70]. Activated Müller glial cells can express glial fibrillary acidic protein (GFAP), an intermediate filament protein, similarly to astrocytes (Fig. 4.4) [71].

2.4 Astrocytes

Another important cell type of the iBRB integrity are the astrocytes. These cells are derived from the visual nerve and spread on the RNFL throughout

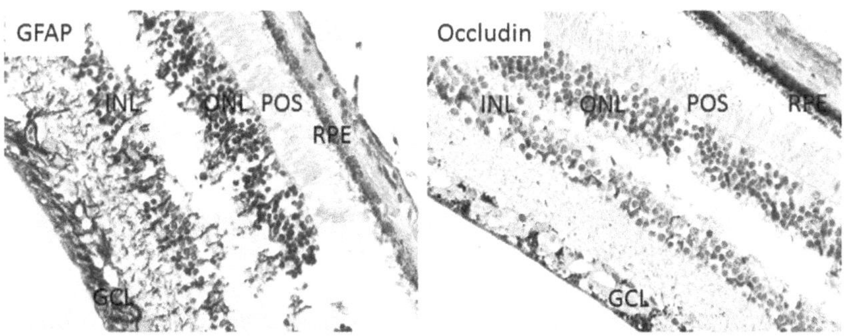

Figure 4.4 Expression of glial fibrillary acidic protein—astrocyte and Müller cell marker (brown; DAB staining [gray in print version]), and occludin (red; StayRed staining [black in print version])—tight junction marker in the human retina. Retinal pigment epithelium (RPE); photoreceptor outer segments (POS); outer nuclear layer (ONL); inner nuclear layer (INL); ganglion cell layer (GCL). Magnification shown: ×63.

development [72,73]. Astrocytes are tightly connected to the capillaries of the retina, similar to pericytes and Müller cells [74,75]. They facilitate production of important molecules needed to form TJ proteins such as ZO-1, claudin-5, occludin, as well as maintain the morphology of endotheliocytes [76,77] (Fig. 4.4). Dysfunction of astrocytes can cause TJ disruption and vascular liquid and protein penetration to the extracellular space, which results in an iBRB integrity disruption and a pathologic condition known as vasogenic edema [78–80].

3. Outer blood-retinal barrier

The transport of molecules through the RPE occurs in transcellular and paracellular manner. In addition, hydrostatic and osmotic pressures are involved in the paracellular transport of little substances, while active transport provides a transcellular motion of molecules [81]. The motion of larger molecules is restricted by the TJ of the RPE [82,83].

The molecular composition of the RPE TJs is similar to those in other tissues and contains occludin-1, claudins, and ZO-1 [82,83]. The actin cytoskeleton in the TJs plays a role of anchor and provides polarization of the RPEs, as well as signal transduction [84,85]. The cell polarity helps localize certain proteins for maintaining the oBRB under normal functioning and helps give appropriate response to environment challenges in the surrounding tissue [86–89]. The TJ can change their permeability under the influence of the RPE or outer retinal stimuli. The RPE produce hepatocyte growth factor (HGF) which brings structural and functional changes of the TJs [90,91]. In addition, HGF overexpression has been shown to cause a chronic retinal detachment (RD) from the BRB under experimental conditions [92].

The external molecular transport through the RPE is to a large extent (70%) dependent upon active ion transport, whereas high-oncotic pressure can influence another 30% of the choroidal molecular transport [93].

RPEs express P-glycoprotein, which facilitates transcellular transport of molecules in these cells, as well as probable elimination of unnecessary metabolites from the subretinal space [94]. The RPEs contain microvilli on the apical side and curved basal membrane that extend the cell surface area and facilitate transcellular transport of molecules in them [95]. The RPE-retina transport of molecules depends on the localization of Na^+-K^+-ATPases found in the apex of the RPE cells and the cytoskeletal proteins found in the inner membrane surface of the RPEs. This protein pattern is opposite to the one found in other epitheliocytes [95]. Despite the sodium and potassium cell ion flux regulation by the Na^+-K^+-ATPases, water flow is not inhibited by ouabain; however, blocking of phosphodiesterase activity can remarkably decrease the water flow [96]. A stable expression of aquaporin (AQ) -1 on the apical membrane of the RPE is also present. The AQ-1 promotes water flow through the RPEs from the subretinal space [97].

RPEs are also responsible for delivery of glucose to the photoreceptor layer of the retina, as well as retinol, which is necessary for the synthesis of visual pigments. Galactose gets captured by the retina and is included into the cones [98]. The glycogen in cones is created with use of galactose [99]. There are nutrient receptors on the basal and lateral cell membranes of the RPE that need to be delivered to the retina. The various retinoid binding proteins in the cytosol of the RPE facilitate the retinol transport [100].

The fenestrated chorio-vessels underneath the RPEs and the Bruch's membrane play an important role in the retinal feeding and in the maintenance of the major basal-apical transport of molecules from the choroid to the retina. The unnecessary metabolites are removed in this manner [3].

3.1 The role of the retinal pigment epithelium in physiology and pathology

The RPEs play an important role in the normal retina functioning and maintains an appropriate work flow in the photoreceptor layer. Furthermore, RPEs take part in the growth factors production, as well as nutrient and metabolite exchange with the photoreceptors. RPEs absorb light and phagocytose photoreceptor outer segments (POSs) [101]. In fact, RPEs phagocytize more material destined to recycling than any other phagocytic cell type in the human body.

Dysfunctions and death of the RPEs can lead to the dramatic worsening of vision, as occurs in certain pathologies of the eye, including age-related macular degeneration (AMD) [102,103], RD, and retinal dystrophies. Structural and molecular changes found in the RPE during disease can help elucidate the origin of the pathology and find ways for improved treatment modalities [102,103].

AMD is the most frequent cause of legal blindness among the elderly in the developed world [102,103]. Although intravitreal injections of VEGF antagonists (anti-VEGF) has been a major breakthrough for the treatment of the neovascular form of AMD, dysfunctional RPE replacement with stem cell transplantation is one of the most encouraging method for general AMD regenerative treatment [104].

Human pluripotent stem cells have endless self-renewal properties [105,106] and have the ability to differentiate into normal functioning RPEs [107]. Extensive efforts throughout the world are conducted to approve stem cell-based treatments in the eye [108–110]. Recently, RPEs derived from human embryonic stem cells (hESC-RPE) generated on a transplantable polyimide membrane coated with a biopolymer [111] and parylene membrane [112] have been shown. The immunogenicity of the hESC-RPE [113], however, needs to be further elucidated, including in vitro purification and death of these cells.

Apoptosis, anoikis, or autophagy-associated cell death are the most common cell death types in the retina, all being described previously as part of the AMD pathogenesis [114,115]. Yellow sediments or drusens are accumulations in the Bruch's membrane and RPE layer interspace, which lead to elevation of the RPEs from the Bruch's membrane, thus inducing a cell death by anoikis or delamination from the ECM [116]. In addition, it has been described that the drusen from AMD patients contains microtubule-associated protein light chain 3 (LC3)—the main autophagic marker, which has provided evidence for increased autophagic activity within aging RPEs [117]. Autophagy is a normal response to lack of nutrients, hypoxia, and oxidative stress, and is observed in low levels in almost all tissues [118], including the retina and the RPEs.

The elimination of dead cells is an important process for retinal homeostasis [119]. In case of incomplete elimination, waste or drusen formation can develop, which is the hallmark of the early- or dry-type of AMD [120]. RPEs can engulf efficiently dying RPE cells (Fig. 4.5A).

Consequently, disruption of the BRB by penetration of the Bruch's membrane with new vessels (neovascularization) can lead to the late- or wet-form of AMD, which is associated with presence of professional phagocytes (macrophages and dendritic cells) [121,122]. During apoptosis, cells expose signal molecules on the outer side of the plasma membrane, such as phosphatidylserine, which are "eat-me" signals for the phagocytes to recognize them and contribute to their clearance [123,124].

Dynamic interrelationship between dying and engulfing cells exists with many overlapping molecular elements which constitute the apopto-phagocytic system. This so called "third synapse" between the dying cells and the phagocyte involves elements that can be grouped into the following functional categories: phagocytosis receptors, cell surface molecules,

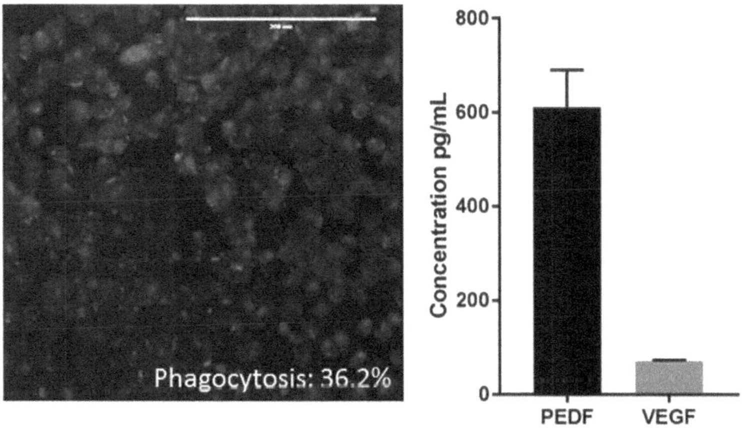

Figure 4.5 Phagocytosis and secretion of pigment epithelium-derived growth factor by ARPE-19 cells: ARPE-19 cells were stained against the actin cytoskeleton (green [dark gray in printed version], Alexa Fluor 488), the engulfed beads (red [light gray in printed version], pHrodo™ Red E. coli BioParticles® Conjugate® (P35361, Thermo Fisher)) and the nuclei (blue [gray in printed version], 4′,6-diamidino-2-phenylindole (DAPI)) are shown (phagocytosis percent represents the ratio of engulfing (green$^+$red$^+$ [dark gray and light gray in printed version]) versus nonengulfing (green$^+$ [dark gray in printed version]) cells over a 2 h phagocytosis time in vitro); scale bar: 200 μm. For the ELISA measurements, 500,000 ARPE-19 cells were seeded for conditioned medium acquisition. Following attachment, cells were incubated for 24 h in basal medium. The media was collected and after removal of cell debris by centrifugation frozen at −20°C, until assayed for PEDF/VEGF secretion by ELISA. The samples were measured in triplicates ($n = 4$). Concentrations of PEDF (611 ± 78.75 pg/mL) and VEGF (70.46 ± 2.64 pg/mL) are shown.

bridging molecules, signal transducers, engulfment proteins, effector molecules, transcription factors, inflammatory regulators, and cytokines. While many redundant elements of the recognition and receptor elements of the apopto-phagocytic machinery have been described in different organ systems, they all seem to converge into the rac-1 dependent cytoskeletal pathway [125].

Corticosteroids (CS) have been extensively used in different systems for their rapid and delayed effects on physiological and pathological functions. In the eye, the CS triamcinolone (TA) is applied intravitreally for its basic antiinflammatory, antipermeability, and antifibrotic properties. Vitreous injection of TC can be a powerful therapeutic agent for treating inflammatory eye diseases as AMD [126−129]. All CSs exert their action via cytosolic CS receptors which move into the nucleus to regulate gene expression or upon ligand binding to membrane-bound CS receptors. The potentiating effect of CS on the phagocytosis of apoptotic neutrophils has been well-described. The enhanced phagocytic uptake of dying cells has been accounted to the increased capacity for engulfment-mediated reorganization of cytoskeletal elements, loss of phosphorylation of adhesion mediators (paxillin and pyk2), and increased amount of Rac GTPase. The effect of TA on the retinal phagocytosis process has been described by our group before [119,125,130,131].

Microarray analysis of the expression patterns of CS-induced human monocytes has found specific gene-clusters as possible functional markers of the anti-inflammatory response: up-regulated antioxidative, migration/chemotaxis, phagocytosis, anti-inflammatory genes, and down-regulated T-cell chemotaxis, adhesion, apoptosis, oxidative functions, and IFNγ regulated genes. The importance of the tyrosine kinase (Tyro3/Axl/Mer) family of receptors in the phagocytosis of apoptotic cells has been clearly demonstrated in dexamethasone treated human blood monocyte derived macrophages (HMDMs), which exhibit augmented capacity for phagocytosis in the presence of protein S or growth-arrest specific 6 (Gas6), known ligands for Mer tyrosine kinase (Mertk). A population-based polymorphism study found an interaction of Gas6 and Mertk with other risk factors in the pathogenesis of AMD [132]. It has been shown that TC treatment of macrophages can lead to increased removal of anoikis RPE in vitro [125]. Furthermore, up-to-date investigations have shown that macrophages secrete interleukin (IL) -6, tumor necrosis factor (TNF) -α and IL-8 cytokines, which appear to affect the phagocytosis of autophagy-associated dying cells [133], as well as angiogenesis and balance of VEGF/PEDF ratio secreted by RPEs (Fig. 4.5B; background expression of PEDF by the RPE cell line ARPE-19).

During AMD pathogenesis, weak chronic inflammation—an extended level of inflammatory mediators is observed. These factors are known to launch and modulate neovascularization, inflammation, and distribution of inflammatory cells [134].

A summary of the apopto-phagocytic synapse processes occurring during dry and wet type of AMD is shown in Fig. 4.6.

3.2 The role of reactive oxygen species in retinal disease pathogenesis

ROS include oxygen free radicals with unpaired electrons—superoxide (O_2-\$), hydroxyl (\$OH), and peroxyl as well as oxidizing agents not considered as free radicals—hydrogen peroxide (H_2O_2), hypochlorous acid, and

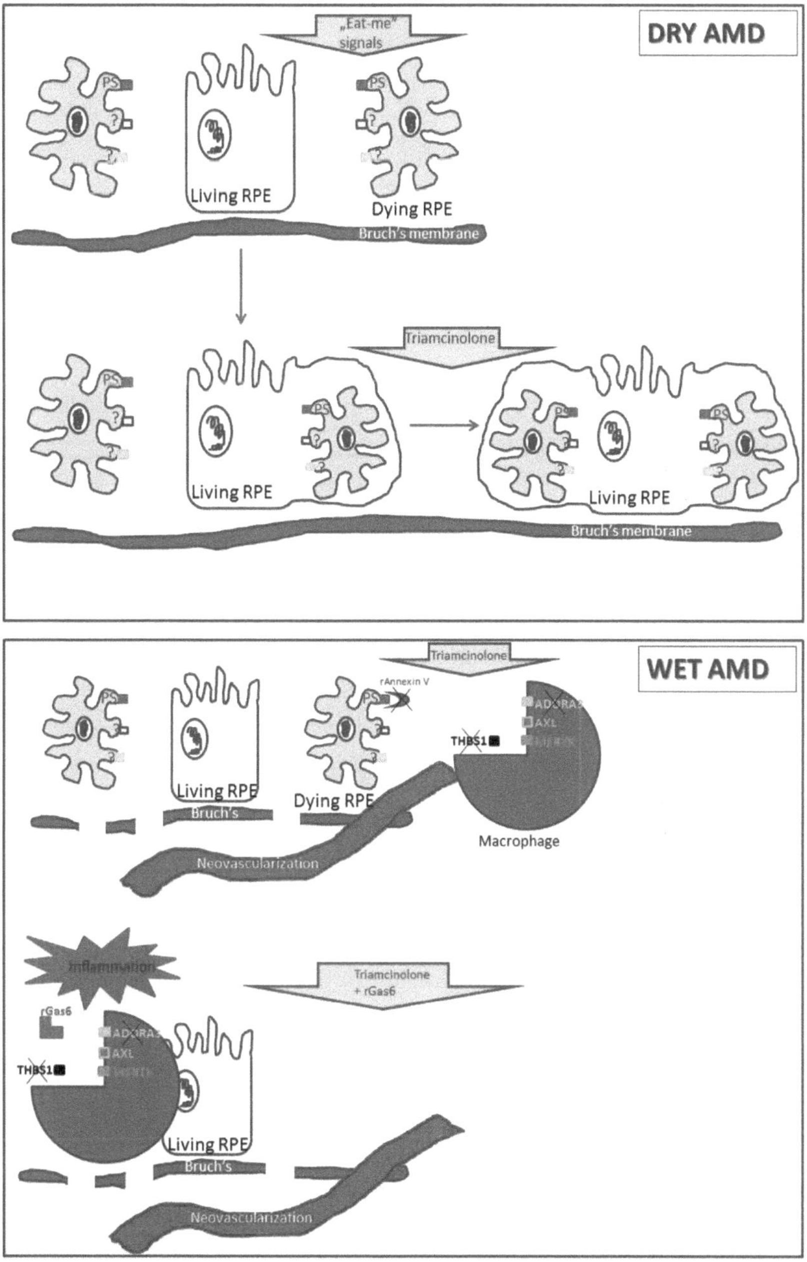

Figure 4.6 Apopto-phagocytic synapse in age-related macular degeneration: During dry AMD, the Bruch's membrane is intact and dying RPE cells which expose phosphatidylserine (PS) "eat-me" signals on their cell membranes get phagocytosed by neighboring living RPEs; this process is facilitated by triamcinolone. In wet AMD, the Bruch's

ozone (O_3) [135]. The sources for ROS can be endogenous or exogenous. The main endogenous source of ROS is the mitochondrial electron transport chain, which can leak electrons to oxygen through a chain of redox reactions [136]. Further endogenous sources include NADPH oxidases (Nox) on phagocytes and endothelial cell membranes [137,138], as well as cytochrome P450 oxygenases, xanthine oxidase, lipoxygenase, and cyclooxygenase [136,139]. Well-known exogenous sources include tobacco smoke, fatty acids in foods, transition metals, and ethanol, all of which can produce hydroxyl and organic radicals as well as lipid peroxidases [139]. Electromagnetic radiations, including gamma rays, X-rays, and ultraviolet radiation, serve as a physical source of ROS [139], while quinones, alkylating agents, aldehydes, and heavy metals can induce production of ROS by altering cellular responses [139].

Different concentrations of ROS have different positive or negative effects at cellular level. In low concentrations, ROS play an essential role in altering protein structure, thereby changing its function. They have been shown to interact with effectors in signal transduction pathways such as ras and protein kinase C [139] and can affect multiple cell signaling pathways involving essential functions such as proliferation, inflammation, apoptosis, and gene expression [135]. Many growth factors, such as platelet-derived growth factor and epidermal growth factor, as well as cytokines and angiotensin II, release a burst of ROS when they are bound to their respective receptors, which is essential for their function [139].

Higher concentrations of ROS, however, are detrimental to health and have been shown to bind to segments of DNA, which can potentially lead to mutations. Although small amounts of damage can usually be repaired, larger amounts of damage can lead to cell death [139]. ROS can also damage the plasma membranes of cells and interrupt the cellular homeostasis by oxidizing in the normally reductive intracellular environment, altering intracellular and intercellular functions [135,139].

It is evident that ROS production needs to be offset by ROS removal for health to be maintained. Enzymatic and nonenzymatic antioxidants exist in

membrane or the outer blood-retinal barrier gets interrupted or penetrated by new vessels from the choroid, which allows professional macrophages to facilitate phagocytosis. The engulfment of dying RPEs is not PS-dependent nor thrombospondin 1 (THBS1)- or adenosine receptor 1 (ADORA1)-dependent. Triamcinolone itself, and recombinant Gas6 (rGas6) further facilitate the engulfment process; however, this is accompanied by release of proinflammatory cytokines (IL-6, IL-8, and TNFα).

cell cytoplasm and mitochondria to minimize ROS related damage. Enzymatic antioxidants such as glutathione peroxidase, catalase, and superoxide dismutase (SOD), and nonenzymatic antioxidants such as glutathione, vitamin C and E, uric acid, and selenium are oxidized by ROS, which is in turn reduced and inactivated [140], keeping the amount of ROS in fine balance within the cell.

ROS are known to be involved in retinal disease pathogenesis. The retina is particularly vulnerable to oxidative damage. It is one of the highest consumers of oxygen in the human body, which serves as a fertile environment for the generation of ROS [141,142]. Retinal cells are also extremely metabolically active and are exposed to high levels of light and other electromagnetic radiation, both of which are a major source of oxidative stress [143], while the BRB is not a true barrier for ROS formation.

The ROS are not only generated in higher amounts in the retina, but the potential for oxidative damage among retinal cells is also increased. Photoreceptors are rich in polyunsaturated fatty acids, particularly docosahexaenoic acid, which is extremely susceptible to lipid peroxidation [144]. To overcome this, the retina possesses multiple ROS scavengers [143] and continually phagocytoses the POS to neutralize oxidative damage [144]. This delicate balance is disrupted in retinal diseases, in particular, diabetic retinopathy, AMD, myopia, retinal vein occlusion, retinitis pigmentosa, and retinopathy of prematurity.

3.3 The role of melanin in retinal pigment epithelium antioxidative processes

The high-oxidative stress RPEs undergo is balanced by the presence of the highly pigmented melanosomes in their cytoplasm. The melanin content acts as an antioxidative or protective molecule binding metals and removing ROS [145].

Ex vivo human RPE cultures show high maturity and differentiation by their presence or absence of melanosomes, as well as by forming their own collagen ECM (Fig. 4.7). The size, density, and shape of the melanosomes is important factor determining the capability of the cells to withstand oxidative stress [146]. The stage these organelles are found in ranges from I—IV—this determines their developmental progression. Stage I (premelanosomes) are spherical structures which originate from the smooth endoplasmic reticulum and contain transmembrane protein, gp 100. Stage II melanosomes are ovoid in shape and contain and internal fibrillar matrix, arising from cleavage

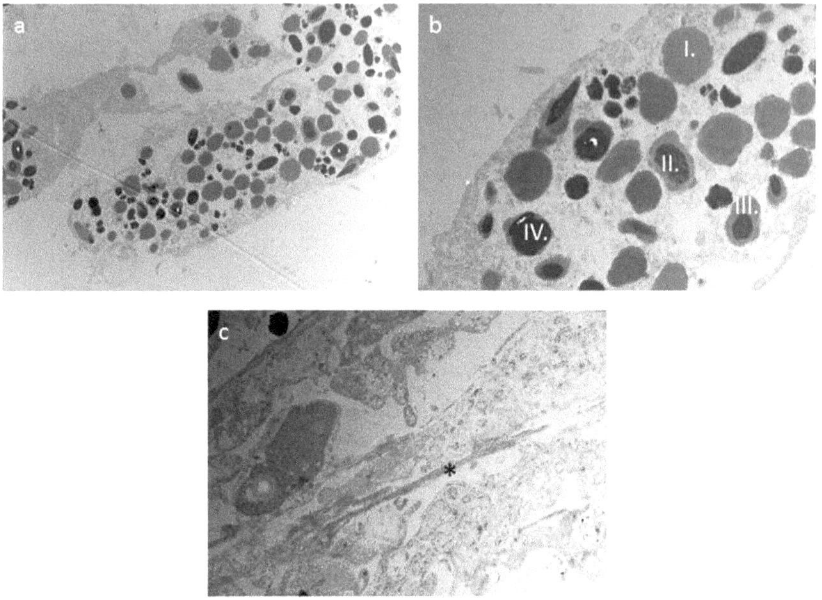

Figure 4.7 Transmission electron microscopy of ex vivo cultivated primary human retinal pigment epitheliums (A) having a high content of melanosomes at different stages of maturations (I–IV) (B), and de novo produced collagen extracellular matrix (*) (C).

of gp100. Stage III melanosomes are the ones in which melanin deposition begins, while Stage IV (mature) melanosomes are filled with melanin [147].

The melanosomes with their melanin content represent a barrier in itself, absorbing the incoming light, thus not allowing it to penetrate further into the choroid.

In albinism, low amount of melanin in melanosomes is because of a defect in melanin biosynthesis; abnormal biogenesis of melanosome OCA1, OCA3; tyrosinase (OCA1) and tyrosinase-related protein (OCA3) retained in rough endoplasmic reticulum. Membrane-associated transporter protein (MATP) interferes with processing and trafficking of tyrosinase to the melanosome and results in secretion of early melanosomes [148].

3.4 Blood-retinal barrier in uveitis

Uveitis is defined as intraocular inflammation of the uveal tract and is clinically classified as anterior, intermediate, posterior, or pan-uveitis depending on the major site of inflammation. Uveitis is vastly heterogenic and complex, and its etiology can be autoimmune or infectious. It is responsible for 10% of blindness in western countries and may affect patients of all ages [149]. Animal models can be useful to study disease pathogenesis, even though these models do not exactly mirror all aspects of human disease. One of the most widely used animal models is experimental autoimmune uveoretinitis (EAU), which can model posterior uveitis [150]. In health, the BRB ensures a form of "immune privilege" which protects the retina from potentially harmful immune responses [151]. However, upon inflammation of the retina (as in uveitis), the BRB becomes compromised and leukocytes cross the BRB and migrate into the retina. Diapedesis (extravasation) of leukocytes across the BRB necessitates, among other things, up-regulation of adhesion molecules. In mouse EAU, leukocyte adhesion is enabled through a number of adhesion molecules, including P- and E-selectins, as wells as ICAM-1, which are expressed on retinal endothelial cells [152]. In line with the latter animal study, elevated concentrations of soluble P- and E-selectin and ICAM-1 have been detected in patients with primary retinal vasculitis, Behcet's disease, sarcoidosis, and idiopathic uveitis [153—158]. In addition to adhesion molecules, chemokines such as CCL2, CCL3, and CCL10, produced by retinal endothelial cells, may attract and support leukocyte tissue infiltration [159,160]. In EAU, leukocyte diapedesis takes place primarily at the postcapillary retinal venules of the iBRB rather than at the oBRB, and mainly by a transcellular route rather than a paracellular route [152,161]. Paracellular migration of leukocytes cannot be completely excluded, however, as leukocyte diapedesis causes disruption of the junctional protein occluding-1 in mice EAU [162]. Whether leukocyte diapedesis is necessary to induce breakdown of the BRB has been a matter of some controversy. In EAU, BRB breakdown, as evidenced by permeability to lanthanum and horseradish peroxidase, was seen only after lymphocyte infiltration of the retina had occurred [161]. However, retinal endothelial cells are also able to produce TNFα and IL-1β which promote breakdown of the TJs of the BRB [163—165]. The recent success of employing TNFα inhibitors in the treatment of uveitis therefore emphasizes the importance of the BRB in uveitis.

3.5 Retinal pigment epithelium tissue engineering and processing of samples

Isolation and cultivation of primary RPEs is important step in cell and tissue engineering. We hereby describe a well-established protocol for such isolation of porcine RPEs.

Freshly enucleated porcine eyeballs are needed. The isolation of primary RPE cells begins by removing excess tissue from the eye and sterilizing in 0.2% Betadine solution for 10 min in ice. Eyes are then rinsed properly in ddH_2O and stored in a solution of Penicillin/Streptomycin on ice. Next, a circumferential section at *ora serrata* is made, and the anterior eye segment and the vitreous are removed. The eye cups are then filled with 1 mM EDTA and incubated at 37°C, 5% CO_2 for 30 min. After incubation, the neural retina is gently loosened with forceps and separated from the optic nerve using scissors. After washing with PBS, the eyecup is filled with 0.25% Trypsin/ EDTA solution and incubated at 37°C for 20 min. After incubation, the trypsin solution is removed and the residual trypsin is inhibited by the addition of culture medium consisting of DMEM/F12 (D6421, Sigma Aldrich), 10% FBS (F6178, Sigma Aldrich), and 2.5 mM GlutaMAX-1 (35050-038, Life Technologies). The cell suspension is centrifuged for 5 min at 300 g at room temperature. Each cell pellet isolated from a single eye is resuspended in culture medium and transferred to each well of a 12-well plate. For the first 24 h, the cells are cultured in DMEM medium with EGF 50 ng/mL, then changed to DMEM medium containing 10 ng/ mL EGF. RPE cells are propagated in the culture medium, which is changed three times a week. A typical culture of porcine primary RPEs and their characteristic cellular markers is shown in Fig. 4.8.

In addition, to characterize cells and tissues containing pigment such as melanin, a bleaching method needs to be applied before fluorescent immunostaining. We hereby describe a well-established protocol for such processing. Standard deparaffinization of the slides to be stained is first performed. After deparaffinization, bleaching is performed by first rinsing the tissue on the slide with PBS, then incubating with KMnO 0.25% for 30 min at room

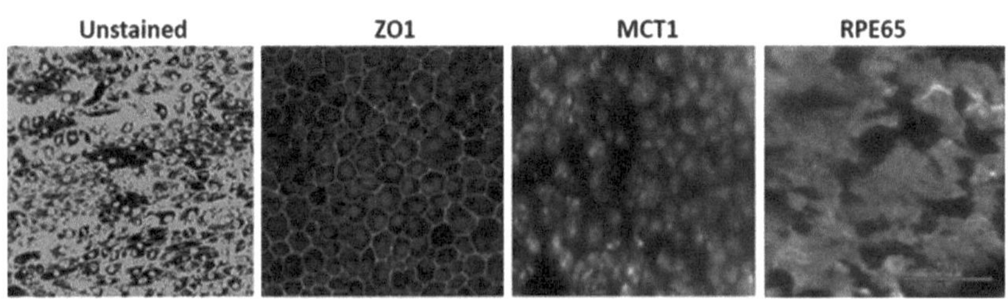

Figure 4.8 Representative images of porcine primary RPEs cultured on glass (unstained) or nanofibrous membrane stained by fluorescent immunocytochemistry. Relatively high presence of melanin in unstained cultured cells and high expression (green staining [gray in print version]) of typical markers for RPE cells such as zonula occludens (ZO-1), monocarboxylate transporter 1 (MCT1), and retinal pigment epithelium-specific 65 kDa protein (RPE65) can be seen.

temperature, followed by one wash with PBS for 5 min. Consequently, incubation with 1% oxalic acid in PBS solution for 10 min at room temperature is performed until the cells are clear. Washing with PBS for 5 min at room temperature is then performed, and quality check of the pigmented cell layers with fluorescence microscopy to determine whether autofluorescence is extinguished. The samples are then boiled for 5 min in citrate buffer (pH 6.0), washed once in PBS for 5 min at room temperature, and permeabilized by 0.5% Triton X-100 for 10 min. After additional washing with PBS and blocking with 5% BSA/PBS for 1 h, the primary antibodies are applied diluted in 1% BSA/PBS to the dilution indicated in the manufacturer's instructions and the slides are incubated overnight at 4°C. Washing three times for 5 min with PBS and preparation of species-specific secondary antibodies diluted in 1% BSA/PBS to the dilution indicated in the manufacturer's instructions and incubation of tissue for 1 h at room temperature covered from direct light, takes place next. This is followed by washing three times for 5 min with PBS, while protecting the slides from direct light. DAPI is optionally used for nucleus counterstaining and/or together with the mounting medium.

3.6 Retinal organoids and three-dimensional engineering—building the future barriers

Since the emergence of embryonic- and induced pluripotent-stem cells (ESC and iPSC)-derived RPEs and the primary RPEs from allogenic or autologous sources, and ongoing effort has been made to engineer retinal cell layers (mainly RPEs) and retinal organoids. All studies so far consider having a tangible GMP-grade product to take into clinical trials, but there is still limited knowledge whether the natural tissue barriers in the eye or the retina can be regenerated or artificially produced [166].

Under appropriate three-dimensional (3D) culture conditions, ESC and iPSC are capable of differentiating into self-organizing 3D retinal tissue with the different major retinal cell types arranged in their proper layers [167—170]. Photoreceptor cells in the stem cell-derived 3D retinal tissues can achieve an advanced degree of maturation including formation of outer segments, expression of phototransduction proteins, and response to light [170].

We have previously shown self-organizing capability of corneal stroma stem cells under long-term cultivation, and formation of ex vivo engineered tissue with cell and ECM characteristics similar to the human tissue [171]. Analogously, decomposition or gentle digestion of the human neuroretina with trypsin/dispase/collagenase cocktail to a single-cell stage and self-organization of the neuroretina over two months can lead to 3D structures containing GFAP positive, and most importantly, rhodopsin positive/photoreceptor cells (Fig. 4.9).

4. Structural changes on optical coherence tomography and fluorescein angiography in the internal blood-retinal barrier and outer blood-retinal barrier in disease states

Common eye disorders associated with vessel leakage, nonperfusion and microvascular abnormalities are diabetes retinopathy and retinal vein occlusion, as well as choroidal neovascularization in AMD and myopia. State-of-the-art imaging technology is used to diagnose and monitor treatment of such diseases (Fig. 4.10). Noninvasive optical coherence tomography (OCT)

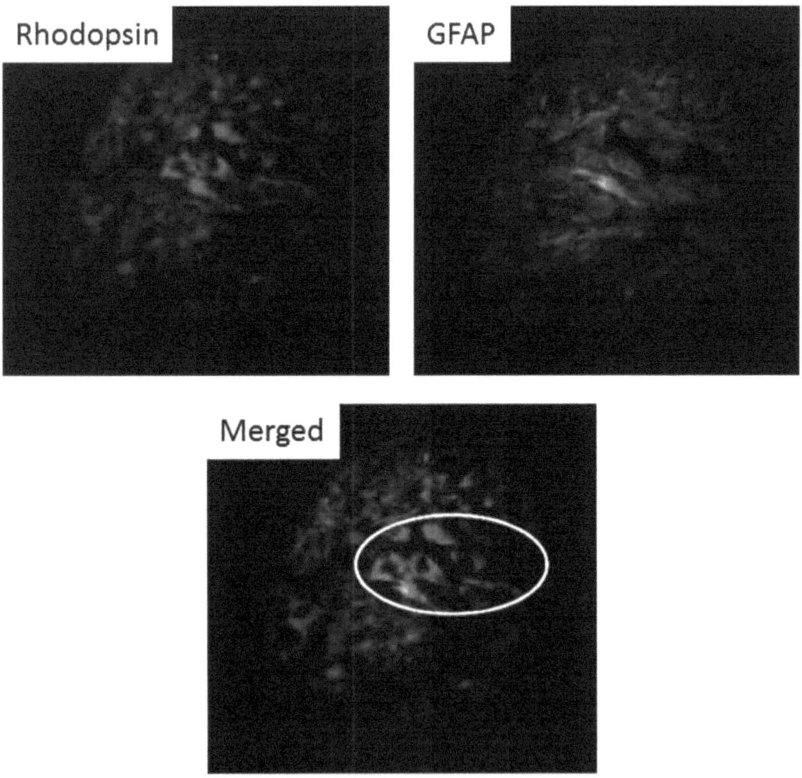

Figure 4.9 Decomposed human neuroretina and recomposed/self-assembled three-dimensional engineered tissue maintained in culture ex vivo in DMEM containing 10% FBS for two months after recomposition. The few rhodopsin positive (red [dark gray in print version]) cells are shown in the encircled are, being surrounded by the GFAP positive (green [gray in print version]) cells.

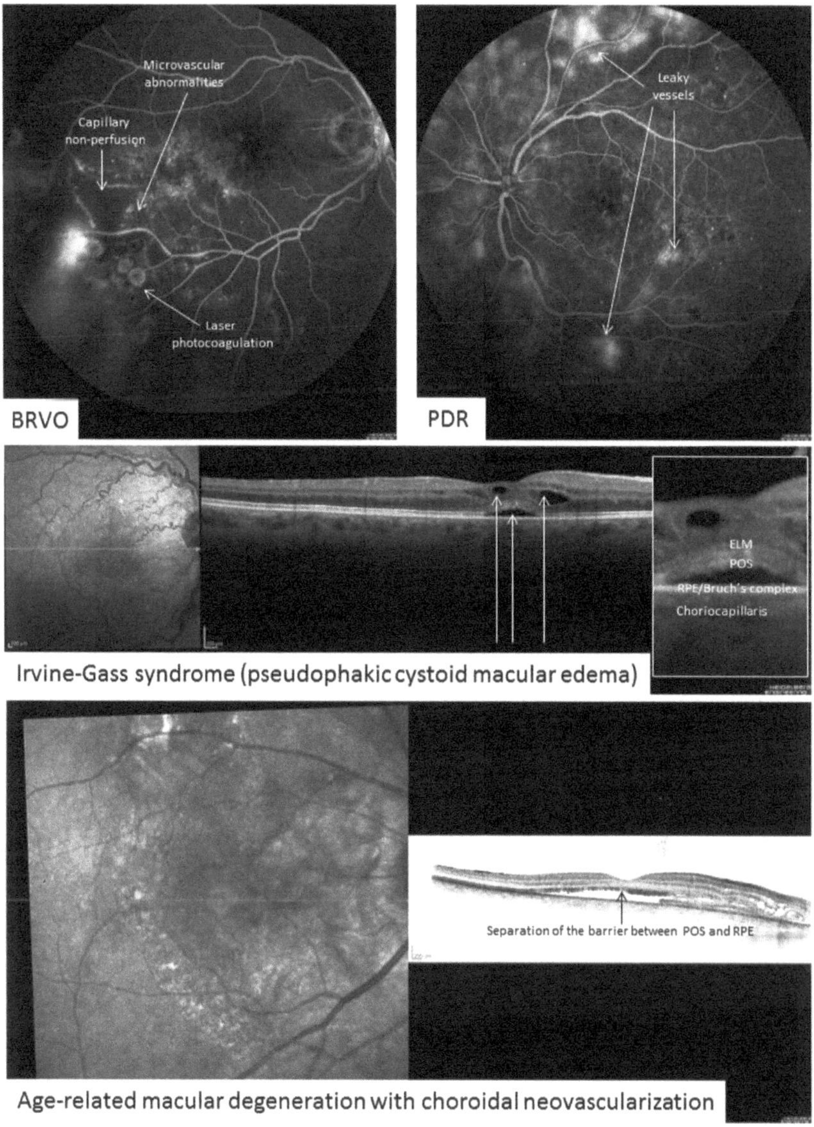

Figure 4.10 Inner and outer blood-retinal barrier changes in branch retinal vein occlusion (BRVO), proliferative diabetic retinopathy (PDR), Irvine-Gass syndrome, and neovascular age-related macular degeneration (wet AMD). Fluorescein angiography fundus images and corresponding optical coherence tomography images are shown. *ELM*, external limiting membrane; *POS*, photoreceptor outer segments; *RPE*, retinal pigment epithelium.

and OCT angiography as well as upcoming new technologies allowing vascular flow analysis in the retina and the choroid are gradually substituting more invasive fluorescein angiography, which is still considered the gold-standard for detecting pathological vessels. Furthermore, novel molecular and structural indicators for detection of changes in the iBRB and oBRB are promising future disease biomarkers.

5. Conclusions

The BRB with its inner and outer part is a tight and restrictive system which functions under normal conditions to regulate ion, protein, and water dynamics in and out of the retina. The complex interplay of cells and TJs between them, in particular, that between vascular endothelial cells and RPEs, are important to maintain the privileged state of the eye compared to other organ systems. The role RPEs play in physiology and disease, phagocytosis and maintaining oxidative balance is essential to characterize in functional studies. Barrier formation should be considered of equal importance as two-dimensional and 3D tissue engineering in future regenerative medicine work. In addition, advancements in noninvasive technologies for studying the BRB and BRB-disruptions in retinal diseases are of vital importance.

Acknowledgments

The authors like to thank Giang H. Nguyen from the Center for Eye Research, Department of Ophthalmology, University of Oslo, and Yiqing Cai from the Institute for Oral Biology, University of Oslo, for their kind help in processing the immunohistochemistry and transmission electron microscopy samples, respectively. This work was funded by the South-Eastern Norway Regional Health Authority (project No. 2017105), the Norwegian Association of the Blind and Partially Sighted, the Czech Science Foundation (Project Number 18-04393S), and the Technology Agency of the Czech Republic (KAPPA Programme, Project Number TO01000107). All authors contributing to the study have read and approved the manuscript. There are no conflicts of interest for any of the authors.

References

[1] Hoon M, Okawa H, Della Santina L, Wong ROL. Functional architecture of the retina: development and disease. Prog Retin Eye Res 2014;42:44–84. https://doi.org/10.1016/j.preteyeres.2014.06.003.

[2] Joyce C, Le PH, Sadiq NM. Retina. In: Histology. StatPearls Publishing; 2019.

[3] Kaur C, Foulds WS, Ling EA. Blood-retinal barrier in hypoxic ischaemic conditions: basic concepts, clinical features and management. Prog Retin Eye Res 2008;27: 622–47. https://doi.org/10.1016/j.preteyeres.2008.09.003.

[4] Shakib M, Cunha-Vaz JG. Studies on the permeability of the blood-retinal barrier. IV. Junctional complexes of the retinal vessels and their role in the permeability of the blood-retinal barrier. Exp Eye Res 1966;5. https://doi.org/10.1016/S0014-4835(66)80011-8.

[5] Ashton N. Bowman lecture. The blood-retinal barrier and vaso-glial relationships in retinal disease. Trans Ophthalmol Soc U K 1965;85:199–230.

[6] Reichenbach A, Wurm A, Pannicke T, et al. Müller cells as players in retinal degeneration and edema. Graefe's Arch Clin Exp Ophthalmol 2007;245:627–36.

[7] Cunha-Vaz JG. The blood-retinal barriers. Doc Ophthalmol 1976;41:287–327. https://doi.org/10.1007/BF00146764.

[8] Henkind P. Symposium on glaucoma: joint meeting with the National Society for the Prevention of Blindness. New observations on the radial peripapillary capillaries. Invest Ophthalmol 1967;6:103–8.

[9] Bernstein MH, Hollenberg MJ. Fine structure of the choriocappillaris and retinal capillaries. Invest Ophthalmol 1965;4:1016–25.

[10] Schnitzer JE, Liu J, Oh P. Endothelial caveolae have the molecular transport machinery for vesicle budding, docking, and fusion including VAMP, NSF, SNAP, annexins, and GTPases. J Biol Chem 1995;270:14399–404. https://doi.org/10.1074/jbc.270.24.14399.

[11] Simionescu M, Gafencu A, Antohe F. Transcytosis of plasma macromolecules in endothelial cells: a cell biological survey. Microsc Res Tech 2002;57:269–88. https://doi.org/10.1002/jemt.10086.

[12] Lossinsky AS, Shivers RR. Structural pathways for macromolecular and cellular transport across the blood-brain barrier during inflammatory conditions. Rev Histol Histopathol 2004;19:535–64.

[13] Grierson I, Lee WR. Pressure effects on flow channels in the lining endothelium of schlemm's canal: a quantitative study by transmission electron microscopy. Acta Ophthalmol 1978;56:935–52. https://doi.org/10.1111/j.1755-3768.1978.tb03813.x.

[14] Sagaties MJ, Raviola G, Schaeffer S, Miller C. The structural basis of the inner blood-retina barrier in the eye of *Macaca mulatta*. Invest Ophthalmol Vis Sci 1987;28. 2000–14.

[15] Vinores SA, Derevjanik NL, Ozaki H, et al. Cellular mechanisms of blood-retinal barrier dysfunction in macular edema. Doc Ophthalmol 1999;97:217–28. https://doi.org/10.1023/A:1002136712070.

[16] Hofman P, Blaauwgeers HGT, Tolentino MJ, et al. VEGF-A induced hyperpermeability of blood-retinal barrier endothelium in vivo is predominantly associated with pinocytotic vesicular transport and not with formation of fenestrations. Curr Eye Res 2000;21:637–45. https://doi.org/10.1076/0271-3683(200008)2121-VFT637.

[17] Grammas P, Riden M. Retinal endothelial cells are more susceptible to oxidative stress and increased permeability than brain-derived endothelial cells. Microvasc Res 2003;65:18–23. https://doi.org/10.1016/S0026-2862(02)00016-X.

[18] van Meer G, Simons K. The function of tight junctions in maintaining differences in lipid composition between the apical and the basolateral cell surface domains of MDCK cells. EMBO J 1986;5:1455–64. https://doi.org/10.1002/j.1460-2075.1986.tb04382.x.

[19] Mandel LJ, Bacallao R, Zampighi G. Uncoupling of the molecular "fence" and paracellular "gate" functions in epithelial tight junctions. Nature 1993;361:552–5. https://doi.org/10.1038/361552a0.

[20] Takano K, Kojima T, Sawada N, Himi T. Role of tight junctions in signal transduction: an update. EXCLI J 2014;13:1145–62.
[21] Osanai M, Takasawa A, Murata M, Sawada N. Claudins in cancer: bench to bedside. Pflugers Arch. Eur. J. Physiol. 2017;469:55–67.
[22] Furuse M, Fujita K, Hiiragi T, et al. Claudin-1 and -2: novel integral membrane proteins localizing at tight junctions with no sequence similarity to occludin. J Cell Biol 1998;141:1539–50. https://doi.org/10.1083/jcb.141.7.1539.
[23] Mitic LL, Anderson JM. Molecular architecture of tight junctions. Annu Rev Physiol 1998;60:121–42. https://doi.org/10.1146/annurev.physiol.60.1.121.
[24] Russ PK, Davidson MK, Hoffman LH, Haselton FR. Partial characterization of the human retinal endothelial cell tight and adherens junction complexes. Invest Ophthalmol Vis Sci 1998;39:2479–85.
[25] Petty MA, Lo EH. Junctional complexes of the blood-brain barrier: permeability changes in neuroinflammation. Prog Neurobiol 2002;68:311–23.
[26] Bazzoni G, Dejana E. Endothelial cell-to-cell junctions: molecular organization and role in vascular homeostasis. Physiol Rev 2004;84:869–901.
[27] Balda MS, Matter K. Tight junctions as regulators of tissue remodelling. Curr Opin Cell Biol 2016;42:94–101.
[28] Dallasta LM, Pisarov LA, Esplen JE, et al. Blood-brain barrier tight junction disruption in human immunodeficiency virus-1 encephalitis. Am J Pathol 1999;155: 1915–27. https://doi.org/10.1016/S0002-9440(10)65511-3.
[29] Jerdan JA, Glaser BM. Retinal microvessel extracellular matrix: an immunofluorescent study. Invest Ophthalmol Vis Sci 1986;27:194–203.
[30] Marshall GE, Konstas AG, Lee WR. Ultrastructural distribution of collagen types I-VI in aging human retinal vessels. Br J Ophthalmol 1990;74:228–32. https://doi.org/10.1136/bjo.74.4.228.
[31] Ponsioen TL, Van Luyn MJA, Van Der Worp RJ, et al. Collagen distribution in the human vitreoretinal interface. Invest Ophthalmol Vis Sci 2008;49:4089–95. https://doi.org/10.1167/iovs.07-1456.
[32] Hirschi KK, D'Amore PA. Pericytes in the microvasculature. Cardiovasc Res 1996; 32:687–98.
[33] Frank RN, Turczyn TJ, Das A. Pericyte coverage of retinal and cerebral capillaries. Invest Ophthalmol Vis Sci 1990;31:999–1007.
[34] Sims D. Recent advances in pericyte biology – implications for health and disease. Can J Cardiol 1991;7:431–43.
[35] Shepro D, Morel NML. Pericyte physiology. Faseb J 1993;7:1031–8.
[36] Balabanov R, Dore-Duffy P. Role of the CNS microvascular pericyte in the blood-brain barrier. J Neurosci Res 1998;53:637–44.
[37] Forbes MS, Rennels ML, Nelson E. Ultrastructure of pericytes in mouse heart. Am J Anat 1977;149:47–69. https://doi.org/10.1002/aja.1001490105.
[38] Wallow IH, Burnside B. Actin filaments in retinal pericytes and endothelial cells. Invest Ophthalmol Vis Sci 1980;19:1433–41.
[39] Matsugi T, Chcn Q, Anderson DR. Contractile responses of cultured bovine retinal pericytes to angiotensin II. Arch Ophthalmol 1997a;115:1281–5. https://doi.org/10.1001/archopht.1997.01100160451011.
[40] Schönfelder U, Hofer A, Paul M, Funk RH. In situ observation of living pericytes in rat retinal capillaries. Microvasc Res 1998;56:22–9. https://doi.org/10.1006/mvre.1998.2086.
[41] Bandopadhyay R, Orte C, Lawrenson JG, et al. Contractile proteins in pericytes at the blood-brain and blood-retinal barriers. J Neurocytol 2001;30:35–44. https://doi.org/10.1023/A:1011965307612.

[42] Haefliger IO, Anderson DR. Oxygen modulation of guanylate cyclase-mediated retinal pericyte relaxations with 3-morpholino-sydnonimine and atrial natriuretic peptide. Invest Ophthalmol Vis Sci 1997;38:1563–8.
[43] Haefliger FO, Zschauer A, Anderson DR. Relaxation of retinal pericyte contractile tone through the nitric oxide-cyclic guanosine monophosphate pathway. Invest Ophthalmol Vis Sci 1994;35:991–7.
[44] Chen Q, Anderson DR. Effect of CO_2 on intracellular pH and contraction of retinal capillary pericytes. Invest Ophthalmol Vis Sci 1997;38:643–51.
[45] Matsugi T, Chen Q, Anderson DR. Adenosine-induced relaxation of cultured bovine retinal pericytes. Invest Ophthalmol Vis Sci 1997b;38:2695–701.
[46] Wang YL, Hui YN, Guo B, Ma JX. Strengthening tight junctions of retinal microvascular endothelial cells by pericytes under normoxia and hypoxia involving angiopoietin-1 signal way. Eye (Lond) 2007;21:1501–10. https://doi.org/10.1038/sj.eye.6702716.
[47] Cogan DG, Toussaint D, Kuwabara T. Retinal vascular patterns: IV. Diabetic retinopathy. Arch Ophthalmol 1961;66:366–78. https://doi.org/10.1001/archopht.1961.00960010368014.
[48] Kuwabara T, Cogan DG. Retinal vascular patterns: VI. Mural cells of the retinal capillaries. Arch Ophthalmol 1963;69:492–502. https://doi.org/10.1001/archopht.1963.00960040498013.
[49] Wallow IHL, Engerman RL. Permeability and patency of retinal blood vessels in experimental diabetes. Invest Ophthalmol Vis Sci 1977;16:447–61.
[50] Sakagami K, Wu DM, Puro DG. Physiology of rat retinal pericytes: modulation of ion channel activity by serum-derived molecules. J Physiol 1999;521(3):637–50. https://doi.org/10.1111/j.1469-7793.1999.00637.x.
[51] Barber AJ, Antonetti DA, Gardner TW. Altered expression of retinal occludin and glial fibrillary acidic protein in experimental diabetes. Invest Ophthalmol Vis Sci 2000;41:3561–8.
[52] Hughes S, Gardiner T, Hu P, et al. Altered pericyte-endothelial relations in the rat retina during aging: implications for vessel stability. Neurobiol Aging 2006;27:1838–47. https://doi.org/10.1016/j.neurobiolaging.2005.10.021.
[53] Bringmann A, Skatchkov SN, Pannicke T, et al. Muller glial cells in anuran retina. Microsc Res Tech 2000;50:384–93. https://doi.org/10.1002/1097-0029(20000901)50:5<384::AID-JEMT7>3.0.CO;2-W.
[54] Tout S, Chan-Ling T, Holländer H, Stone J. The role of Müller cells in the formation of the blood-retinal barrier. Neuroscience 1993;55:291–301. https://doi.org/10.1016/0306-4522(93)90473-s.
[55] Distler C, Dreher Z. Glia cells of the monkey retina-II. Muller cells. Vision Res 1996;36:2381–94. https://doi.org/10.1016/0042-6989(96)00005-3.
[56] Penn JS, Madan A, Caldwell RB, et al. Vascular endothelial growth factor in eye disease. Prog Retin Eye Res 2008;27:331–71.
[57] Eichler W, Yafai Y, Kuhrt H, et al. Hypoxia: modulation of endothelial cell proliferation by soluble factors released by retinal cells. Neuroreport 2001;12:4103–8. https://doi.org/10.1097/00001756-200112210-00048.
[58] Eichler W, Kuhrt H, Hoffmann S, et al. VEGF release by retinal glia depends on both oxygen and glucose supply. Neuroreport 2000;11:3533–7. https://doi.org/10.1097/00001756-200011090-00026.
[59] Duh EJ, Yang HS, Suzuma I, et al. Pigment epithelium-derived factor suppresses ischemia-induced retinal neovascularization and VEGF-induced migration and growth. Invest Ophthalmol Vis Sci 2002;43:821–9.

[60] Tretiach M, Madigan MC, Wen L, Gillies MC. Effect of Müller cell co-culture on in vitro permeability of bovine retinal vascular endothelium in normoxic and hypoxic conditions. Neurosci Lett 2005;378:160–5. https://doi.org/10.1016/j.neulet.2004.12.026.
[61] Kaur C, Sivakumar V, Foulds WS. Early response of neurons and glial cells to hypoxia in the retina. Invest Ophthalmol Vis Sci 2006;47:1126–41. https://doi.org/10.1167/iovs.05-0518.
[62] Behzadian MA, Wang XL, Windsor LJ, et al. TGF-beta increases retinal endothelial cell permeability by increasing MMP-9: possible role of glial cells in endothelial barrier function. Invest Ophthalmol Vis Sci 2001;42:853–9.
[63] Giebel SJ, Menicucci G, McGuire PG, Das A. Matrix metalloproteinases in early diabetic retinopathy and their role in alternation of the blood-retinal barrier. Lab Invest 2005;85:597–607. https://doi.org/10.1038/labinvest.3700251.
[64] Jablonski MM, Tombran-Tink J, Mrazek DA, Iannaccone A. Pigment epithelium-derived factor supports normal Müller cell development and glutamine synthetase expression after removal of the retinal pigment epithelium. Glia 2001;35:14–25. https://doi.org/10.1002/glia.1066.
[65] Poitry-Yamate CL, Poitry S, Tsacopoulos M. Lactate released by Muller glial cells is metabolized by photoreceptors from mammalian retina. J Neurosci 1995;15:5179–91. https://doi.org/10.1523/jneurosci.15-07-05179.1995.
[66] Dubois-Dauphin M, Poitry-Yamate C, de Bilbao F, et al. Early postnatal Müller cell death leads to retinal but not optic nerve degeneration in NSE-Hu-Bcl-2 transgenic mice. Neuroscience 2000;95:9–21. https://doi.org/10.1016/s0306-4522(99)00313-9.
[67] Winkler BS, Arnold MJ, Brassell MA, Puro DG. Energy metabolism in human retinal Muller cells. Invest Ophthalmol Vis Sci 2000;41:3183–90.
[68] Foulds WS. Experimental detachment of the retina. Mod Probl Ophthalmol 1961;4:233–5.
[69] Foulds WS, Ikeda H. The effects of detachment of the retina on the induced and resting ocular potentials in the rabbit. Invest Ophthalmol Vis Sci 1966;5:93–108.
[70] Kroll AJ, Machemer R. Experimental retinal detachment in the owl monkey. III. Electron microscopy of retina and pigment epithelium. Am J Ophthalmol 1968;66:410–27. https://doi.org/10.1016/0002-9394(68)91524-9.
[71] Kaur C, Sivakumar V, Yong Z, et al. Blood-retinal barrier disruption and ultrastructural changes in the hypoxic retina in adult rats: the beneficial effect of melatonin administration. J Pathol 2007;212:429–39. https://doi.org/10.1002/path.2195.
[72] Watanabe T, Raff MC. Retinal astrocytes are immigrants from the optic nerve. Nature 1988;332:834–7. https://doi.org/10.1038/332834a0.
[73] Zhang Y, Stone J. Role of astrocytes in the control of developing retinal vessels. Invest Ophthalmol Vis Sci 1997;38:1653–66.
[74] Schnitzer J. Retinal astrocytes: their restriction to vascularized parts of the mammalian retina. Neurosci Lett 1987;78:29–34. https://doi.org/10.1016/0304-3940(87)90556-8.
[75] Schnitzer J. Astrocytes in the Guinea pig, horse, and monkey retina: their occurrence coincides with the presence of blood vessels. Glia 1988;1:74–89. https://doi.org/10.1002/glia.440010109.
[76] Gardner TW. Histamine, ZO-1 and increased blood-retinal barrier permeability in diabetic retinopathy. In: Transactions of the American ophthalmological society; 1995. p. 583–621.
[77] Willis CL, Leach L, Clarke GJ, et al. Reversible disruption of tight junction complexes in the rat blood-brain barrier, following transitory focal astrocyte loss. Glia 2004;48:1–13. https://doi.org/10.1002/glia.20049.

[78] Chan-Ling T, Stone J. Degeneration of astrocytes in feline retinopathy of prematurity causes failure of the blood-retinal barrier. Invest Ophthalmol Vis Sci 1992;33: 2148–59.

[79] Gardner TW, Lieth E, Khin SA, et al. Astrocytes increase barrier properties and ZO-1 expression in retinal vascular endothelial cells. Invest Ophthalmol Vis Sci 1997;38: 2423–7.

[80] Rungger-Brändle E, Dosso AA, Leuenberger PM. Glial reactivity, an early feature of diabetic retinopathy. Invest Ophthalmol Vis Sci 2000;41:1971–80.

[81] Pederson JE. Fluid physiology of the subretinal space. In: Retina. 4th ed. Elsevier Inc.; 2005. p. 1909–20.

[82] Williams CD, Rizzolo LJ. Remodeling of junctional complexes during the development of the outer blood-retinal barrier. Anat Rec 1997;249:380–8. https://doi.org/10.1002/(SICI)1097-0185(199711)249:3<380::AID-AR9>3.0.CO;2-Y.

[83] Tserentsoodol N, Shin BC, Suzuki T, Takata K. Colocalization of tight junction proteins, occludin and ZO-1, and glucose transporter GLUT1 in cells of the blood-ocular barrier in the mouse eye. Histochem Cell Biol 1998;110:543–51. https://doi.org/10.1007/s004180050316.

[84] Tsukita S, Furuse M, Itoh M. Multifunctional strands in tight junctions. Nat Rev Mol Cell Biol 2001;2:285–93.

[85] Aijaz S, Balda MS, Matter K. Tight junctions: molecular architecture and function. Int Rev Cytol 2006;248:261–98.

[86] Rodriguez-Boulan E, Nelson WJ. Morphogenesis of the polarized epithelial cell phenotype. Science 1989;80(245):718–25. https://doi.org/10.1126/science.2672330.

[87] Eaton S, Simons K. Apical, basal, and lateral cues for epithelial polarization. Cell 1995; 82:5–8.

[88] Drubin DG, Nelson WJ. Origins of cell polarity. Cell 1996;84:335–44.

[89] Rizzolo LJ. Polarity and the development of the outer blood-retinal barrier. Histol Histopathol 1997;12:1057–67.

[90] Balda MS, González-Mariscal L, Contreras RG, et al. Assembly and sealing of tight junctions: possible participation of G-proteins, phospholipase C, protein kinase C and calmodulin. J Membr Biol 1991;122:193–202. https://doi.org/10.1007/BF01871420.

[91] Jin M, Barron E, He S, et al. Regulation of RPE intercellular junction integrity and function by hepatocyte growth factor. Invest Ophthalmol Vis Sci 2002;43:2782–90.

[92] Jin M, Chen Y, He S, et al. Hepatocyte growth factor and its role in the pathogenesis of retinal detachment. Invest Ophthalmol Vis Sci 2004;45:323–9. https://doi.org/10.1167/iovs.03-0355.

[93] Negi A, Marmor MF. Quantitative estimation of metabolic transport of subretinal fluid. Invest Ophthalmol Vis Sci 1986;27:1564–8.

[94] Kennedy BG, Mangini NJ. P-glycoprotein expression in human retinal pigment epithelium. Mol Vis 2002;8:422–30.

[95] Gundersen D, Orlowski J, Rodriguez-Boulan E. Apical polarity of Na,K-ATPase in retinal pigment epithelium is linked to a reversal of the ankyrin-fodrin submembrane cytoskeleton. J Cell Biol 1991;112:863–72. https://doi.org/10.1083/jcb.112.5.863.

[96] Miller SS, Hughes BA, Machen TE. Fluid transport across retinal pigment epithelium is inhibited by cyclic AMP. Proc Natl Acad Sci U S A 1982;79:2111–5. https://doi.org/10.1073/pnas.79.6.2111.

[97] Stamer WD, Bok D, Hu J, et al. Aquaporin-1 channels in human retinal pigment epithelium: role in transepithelial water movement. Invest Ophthalmol Vis Sci 2003;44. https://doi.org/10.1167/iovs.03-0001. 2803–8.

[98] Keegan WA, McKechnie NM, Converse CA, Foulds WS. D-[3H]-galactose incorporation in the bovine retina: specific uptake and transport of the radiolabel in cones. Exp Eye Res 1985;40:619–28. https://doi.org/10.1016/0014-4835(85)90084-3.

[99] Shallal A, McKechnie NM, Converse CA, Foulds WS. D-[3H]galactose incorporation into glycogen in retinal cone cells. Exp Eye Res 1990;50:281–8. https://doi.org/10.1016/0014-4835(90)90212-d.

[100] Chader GJ. Retinoids in ocular tissues: binding proteins, transport, and mechanism of action. In: Cell biology of the eye. Elsevier; 1982. p. 377–433.

[101] Strauss O. The retinal pigment epithelium in visual function. Physiol Rev 2005;85: 845–81.

[102] Gehrs KM, Anderson DH, Johnson LV, Hageman GS. Age-related macular degeneration – emerging pathogenetic and therapeutic concepts. Ann Med 2006;38: 450–71.

[103] Klein R, Klein BEK. The prevalence of age-related eye diseases and visual impairment in aging: current estimates. Invest Ophthalmol Vis Sci 2013;54. https://doi.org/10.1167/iovs.13-12789.

[104] Dang Y, Zhang C, Zhu Y. Stem cell therapies for age-related macular degeneration: the past, Present, and future. Clin Interv Aging 2015;10:255–64.

[105] Takahashi K, Tanabe K, Ohnuki M, et al. Induction of pluripotent stem cells from adult human fibroblasts by defined factors. Cell 2007;131:861–72. https://doi.org/10.1016/j.cell.2007.11.019.

[106] Yu J, Vodyanik MA, Smuga-Otto K, et al. Induced pluripotent stem cell lines derived from human somatic cells. Science 2007;80(318):1917–20. https://doi.org/10.1126/science.1151526.

[107] Lu B, Malcuit C, Wang S, et al. Long-term safety and function of RPE from human embryonic stem cells in preclinical models of macular degeneration. Stem Cells 2009; 27:2126–35. https://doi.org/10.1002/stem.149.

[108] Schwartz SD, Hubschman JP, Heilwell G, et al. Embryonic stem cell trials for macular degeneration: a preliminary report. Lancet 2012;379:713–20. https://doi.org/10.1016/S0140-6736(12)60028-2.

[109] Schwartz SD, Regillo CD, Lam BL, et al. Human embryonic stem cell-derived retinal pigment epithelium in patients with age-related macular degeneration and Stargardt's macular dystrophy: follow-up of two open-label phase 1/2 studies. Lancet 2015;385: 509–16. https://doi.org/10.1016/S0140-6736(14)61376-3.

[110] Kamao H, Mandai M, Okamoto S, et al. Characterization of human induced pluripotent stem cell-derived retinal pigment epithelium cell sheets aiming for clinical application. Stem Cell Rep 2014;2:205–18. https://doi.org/10.1016/j.stemcr.2013.12.007.

[111] Subrizi A, Hiidenmaa H, Ilmarinen T, et al. Generation of hESC-derived retinal pigment epithelium on biopolymer coated polyimide membranes. Biomaterials 2012;33:8047–54. https://doi.org/10.1016/j.biomaterials.2012.07.033.

[112] Diniz B, Thomas P, Thomas B, et al. Subretinal implantation of retinal pigment epithelial cells derived from human embryonic stem cells: improved survival when implanted as a monolayer. Invest Ophthalmol Vis Sci 2013;54:5087–96. https://doi.org/10.1167/iovs.12-11239.

[113] Zhao T, Zhang ZN, Westenskow PD, et al. Humanized mice reveal differential immunogenicity of cells derived from autologous induced pluripotent stem cells. Cell Stem Cell 2015;17:353–9. https://doi.org/10.1016/j.stem.2015.07.021.

[114] Kaarniranta K, Sinha D, Blasiak J, et al. Autophagy and heterophagy dysregulation leads to retinal pigment epithelium dysfunction and development of age-related macular degeneration. Autophagy 2013;9:973–84.

[115] Galluzzi L, Vitale I, Aaronson SA, et al. Molecular mechanisms of cell death: recommendations of the nomenclature committee on cell death 2018. Cell Death Differ 2018;25:486−541.

[116] Gilmore AP. Anoikis. Cell Death Differ 2005;12:1473−7. https://doi.org/10.1038/sj.cdd.4401723.

[117] Wang AL, Lukas TJ, Yuan M, et al. Autophagy and exosomes in the aged retinal pigment epithelium: possible relevance to drusen formation and age-related macular degeneration. PLoS One 2009;4:e4160. https://doi.org/10.1371/journal.pone.0004160.

[118] Levine B, Yuan J. Autophagy in cell death: an innocent convict? J Clin Invest 2005;115:2679−88.

[119] Májai G, Petrovski G, Fésüs L. Inflammation and the apopto-phagocytic system. In: Immunology letters; 2006. p. 94−101.

[120] Bowes Rickman C, Farsiu S, Toth CA, Klingeborn M. Dry age-related macular degeneration: mechanisms, therapeutic targets, and imaging. Invest Ophthalmol Vis Sci 2013;54. https://doi.org/10.1167/iovs.13-12757.

[121] Kulkarni AD, Kuppermann BD. Wet age-related macular degeneration. Adv Drug Deliv Rev 2005;57:1994−2009.

[122] Smith AG, Kaiser PK. Emerging treatments for wet age-related macular degeneration. Expet Opin Emerg Drugs 2014;19:157−64.

[123] Elliott MR, Ravichandran KS. Clearance of apoptotic cells: implications in health and disease. J Cell Biol 2010;189:1059−70.

[124] Ravichandran KS. Find-me and eat-me signals in apoptotic cell clearance: progress and conundrums. J Exp Med 2010;207:1807−17.

[125] Albert R, Kristóf E, Zahuczky G, et al. Triamcinolone regulated apopto-phagocytic gene expression patterns in the clearance of dying retinal pigment epithelial cells. A key role of Mertk in the enhanced phagocytosis. Biochim Biophys Acta Gen Subj 2015;1850:435−46. https://doi.org/10.1016/j.bbagen.2014.10.026.

[126] Danis RP, Bingaman DP, Yang Y, Ladd B. Inhibition of preretinal and optic nerve head neovascularization in pigs by intravitreal triamcinolone acetonide. Ophthalmology 1996;103:2099−104. https://doi.org/10.1016/S0161-6420(96)30383-7.

[127] Jonas JB, Kreissig I, Hugger P, et al. Intravitreal triamcinolone acetonide for exudative age related macular degeneration. Br J Ophthalmol 2003;87:462−8. https://doi.org/10.1136/bjo.87.4.462.

[128] Jonas JB, Kreissig I, Degenring RF. Factors influencing visual acuity after intravitreal triamcinolone acetonide as treatment of exudative age related macular degeneration. Br J Ophthalmol 2004;88:1557−62. https://doi.org/10.1136/bjo.2003.039552.

[129] Lim JI, Niec M, Wong V. One year results of a phase 1 study of the safety and tolerability of combination therapy using sustained release intravitreal triamcinolone acetonide and ranibizumab for subfoveal neovascular AMD. Br J Ophthalmol 2015;99:618−23. https://doi.org/10.1136/bjophthalmol-2014-306002.

[130] Petrovski G, Zahuczky G, Katona K, et al. Clearance of dying autophagic cells of different origin by professional and non-professional phagocytes. Cell Death Differ 2007a;14:1117−28. https://doi.org/10.1038/sj.cdd.4402112.

[131] Petrovski G, Berényi E, Moe MC, et al. Clearance of dying ARPE-19 cells by professional and nonprofessional phagocytes in vitro- implications for age-related macular degeneration (AMD). Acta Ophthalmol 2011;89. https://doi.org/10.1111/j.1755-3768.2010.02047.x.

[132] Losonczy G, Vajas A, Takács L, et al. Effect of the Gas6 c.834+7G>A polymorphism and the interaction of known risk factors on AMD pathogenesis in Hungarian patients. PLoS One 2012;7. https://doi.org/10.1371/journal.pone.0050181.

[133] Petrovski G, Zahuczky G, Májai G, Fésüs L. Phagocytosis of cells dying through autophagy evokes a pro-inflammatory response in macrophages. Autophagy 2007b; 3:509–11.
[134] Stanton CM, Wright AF. Inflammatory biomarkers for AMD. Adv Exp Med Biol 2014;801:251–7. https://doi.org/10.1007/978-1-4614-3209-8_32.
[135] Ung L, Pattamatta U, Carnt N, et al. Oxidative stress and reactive oxygen species: a review of their role in ocular disease. Clin Sci 2017;131:2865–83.
[136] Turrens J. Mitochondrial formation of reactive oxygen species. J Physiol 2003;552: 335–44.
[137] Babior BM. The NADPH oxidase of endothelial cells. IUBMB Life 2000;50:267–9.
[138] Babior BM, Lambert JD, Nauseef W. The neutrophil NADPH oxidase. Arch Biochem Biophys 2002;397:342–4.
[139] Martin KR, Barrett JC. Reactive oxygen species as double-edged swords in cellular processes: low-dose cell signaling versus high-dose toxicity. Hum Exp Toxicol 2002; 21:71–5.
[140] Gate L, Paul J, Ba GN, et al. Oxidative stress induced in pathologies: the role of antioxidants. Biomed Pharmacother 1999;53:169–80.
[141] Beatty S, Koh HH, Phil M, et al. The role of oxidative stress in the pathogenesis of age-related macular degeneration. Surv Ophthalmol 2000;45:115–34.
[142] Jarrett SG, Boulton ME. Consequences of oxidative stress in age-related macular degeneration. Mol Aspect Med 2012;33:399–417.
[143] Kortuem K, Geiger LK, Levin LA. Differential susceptibility of retinal ganglion cells to reactive oxygen species. Invest Ophthalmol Vis Sci 2000;41:3176–82.
[144] van Reyk DM, Gillies MC, Davies MJ. The retina: oxidative stress and diabetes. Redox Rep 2003;8:187–92.
[145] Wang Z, Dillon J, Gaillard ER. Antioxidant properties of melanin in retinal pigment epithelial cells. Photochem Photobiol 2006;82:474. https://doi.org/10.1562/2005-10-21-ra-725.
[146] Rachel RA, Nagashima K, O'Sullivan TN, et al. Melanoregulin, product of the dsu locus, links the BLOC-pathway and OA1 in organelle biogenesis. PLoS One 2012;7: e42446. https://doi.org/10.1371/journal.pone.0042446.
[147] Raposo G, Marks MS. The dark side of lysosome-related organelles: specialization of the endocytic pathway for melanosome biogenesis. Traffic 2002;3:237–48.
[148] Cox GF, Fulton AB. Albinism. In: Ocular disease: mechanisms and management. Elsevier Inc.; 2010. p. 461–71.
[149] Suttorp-Schulten MS, Rothova A. The possible impact of uveitis in blindness: a literature survey. Br J Ophthalmol 1996;80:844–8. https://doi.org/10.1136/bjo.80.9.844.
[150] Caspi RR, Roberge FG, Chan CC, et al. A new model of autoimmune disease. Experimental autoimmune uveoretinitis induced in mice with two different retinal antigens. J Immunol 1988;140:1490–5.
[151] Crane IJ, Liversidge J. Mechanisms of leukocyte migration across the blood-retina barrier. Semin Immunopathol 2008;30:165–77. https://doi.org/10.1007/s00281-008-0106-7.
[152] Xu H, Forrester JV, Liversidge J, Crane IJ. Leukocyte trafficking in experimental autoimmune uveitis: breakdown of blood-retinal barrier and upregulation of cellular adhesion molecules. Invest Ophthalmol Vis Sci 2003;44:226–34. https://doi.org/10.1167/iovs.01-1202.
[153] Arocker-Mettinger E, Steurer-Georgiew L, Steurer M, et al. Circulating ICAM-1 levels in serum of uveitis patients. Curr Eye Res 1992;11(Suppl. l):161–6. https://doi.org/10.3109/02713689208999527.

[154] Zaman AG, Edelsten C, Stanford MR, et al. Soluble intercellular adhesion molecule-1 (sICAM-1) as a marker of disease relapse in idiopathic uveoretinitis. Clin Exp Immunol 1994;95:60–5. https://doi.org/10.1111/j.1365-2249.1994.tb06015.x.

[155] Aydintug AO, Tokgoz G, Ozoran K, et al. Elevated levels of soluble intercellular adhesion molecule-1 correlate with disease activity in Behcet's disease. Rheumatol Int 1995;15:75–8. https://doi.org/10.1007/bf00262712.

[156] Haznedaroglu F, Karaaslan Y, Buyukasik Y, et al. Selectin adhesion molecules in Behcet's disease. Ann Rheum Dis 2000;59:61–3. https://doi.org/10.1136/ard.59.1.61.

[157] Sari RA, Kiziltunc A, Taysi S, et al. Levels of soluble E-selectin in patients with active Behcet's disease. Clin Rheumatol 2005;24:55–9. https://doi.org/10.1007/s10067-004-0982-8.

[158] Lee MT, Hooper LC, Kump L, et al. Interferon-beta and adhesion molecules (E-selectin and s-intracellular adhesion molecule-1) are detected in sera from patients with retinal vasculitis and are induced in retinal vascular endothelial cells by Toll-like receptor 3 signalling. Clin Exp Immunol 2007;147:71–80. https://doi.org/10.1111/j.1365-2249.2006.03253.x.

[159] Crane IJ, McKillop-Smith S, Wallace CA, et al. Expression of the chemokines MIP-1alpha, MCP-1, and RANTES in experimental autoimmune uveitis. Invest Ophthalmol Vis Sci 2001;42:1547–52.

[160] Keino H, Takeuchi M, Kezuka T, et al. Chemokine and chemokine receptor expression during experimental autoimmune uveoretinitis in mice. Graefes Arch Clin Exp Ophthalmol 2003;241:111–5. https://doi.org/10.1007/s00417-002-0556-x.

[161] Greenwood J, Howes R, Lightman S. The blood-retinal barrier in experimental autoimmune uveoretinitis. Leukocyte interactions and functional damage. Lab Invest 1994;70:39–52.

[162] Xu H, Dawson R, Crane IJ, Liversidge J. Leukocyte diapedesis in vivo induces transient loss of tight junction protein at the blood-retina barrier. Invest Ophthalmol Vis Sci 2005;46:2487–94. https://doi.org/10.1167/iovs.04-1333.

[163] Luna JD, Chan CC, Derevjanik NL, et al. Blood-retinal barrier (BRB) breakdown in experimental autoimmune uveoretinitis: comparison with vascular endothelial growth factor, tumor necrosis factor alpha, and interleukin-1beta-mediated breakdown. J Neurosci Res 1997;49:268–80. https://doi.org/10.1002/(sici)1097-4547(19970801)49:3<268::aid-jnr2>3.0.co;2-a.

[164] Tezel G, Li LY, Patil RV, Wax MB. TNF-alpha and TNF-alpha receptor-1 in the retina of normal and glaucomatous eyes. Invest Ophthalmol Vis Sci 2001;42:1787–94.

[165] Smith JR, Choi D, Chipps TJ, et al. Unique gene expression profiles of donor-matched human retinal and choroidal vascular endothelial cells. Invest Ophthalmol Vis Sci 2007;48:2676–84. https://doi.org/10.1167/iovs.06-0598.

[166] Singh MS, Park SS, Albini TA, et al. Retinal stem cell transplantation: balancing safety and potential. Prog Retin Eye Res 2020;75:100779. https://doi.org/10.1016/j.preteyeres.2019.100779. 100779.

[167] Meyer JS, Howden SE, Wallace KA, et al. Optic vesicle-like structures derived from human pluripotent stem cells facilitate a customized approach to retinal disease treatment. Stem Cells 2011;29:1206–18. https://doi.org/10.1002/stem.674.

[168] Phillips MJ, Wallace KA, Dickerson SJ, et al. Blood-derived human iPS cells generate optic vesicle-like structures with the capacity to form retinal laminae and develop synapses. Invest Ophthalmol Vis Sci 2012;53:2007–19. https://doi.org/10.1167/iovs.11-9313.

[169] Nakano T, Ando S, Takata N, et al. Self-formation of optic cups and storable stratified neural retina from human ESCs. Cell Stem Cell 2012;10:771–85. https://doi.org/10.1016/j.stem.2012.05.009.

[170] Zhong X, Gutierrez C, Xue T, et al. Generation of three-dimensional retinal tissue with functional photoreceptors from human iPSCs. Nat Commun 2014;5:4047. https://doi.org/10.1038/ncomms5047.

[171] Szabó DJ, Nagymihály R, Veréb Z, et al. Ex vivo 3D human corneal stroma model for schnyder corneal dystrophy – role of autophagy in its pathogenesis and resolution. Histol Histopathol 2018;33:455–62. https://doi.org/10.14670/HH-11-928.

CHAPTER FIVE

Appressorium morphogenesis and penetration in rice blast fungus

Huan-Bin Shi[1,3], Xue-Ming Zhu[1], Yun-Ran Zhang[2], Xiao-Hong Liu[2], Fu-Cheng Lin[1,2]

[1]State Key Laboratory for Managing Biotic and Chemical Treats to the Quality and Safety of Agro-products, Institute of Plant Protection and Microbiology, Zhejiang Academy of Agricultural Sciences, Hangzhou, Zhejiang, China

[2]State Key Laboratory of Rice Biology, Institute of Biotechnology, Zhejiang University, Hangzhou, Zhejiang, China

[3]State Key Laboratory of Rice Biology, China National Rice Research Institute, Hangzhou, Zhejiang, China

1. Introduction

Structural and immune barriers sustain the cellular and molecular integrity and endurance of the host systems. The pathogenicity of microorganisms depends on the efficacy of their interactive mechanisms in breaching host defense and resistance to the infection. In conjunction with the above, investigation of the pathogenicity-driving pathways in fungus-plant models can shed light on the molecular evolution of host-pathogen interactions. Evolutionarily the pathogenic fungi acquired diverse mechanisms to subvert the plant tissue barriers; one of them is emergency of appressorium, a specialized infection structure [1]. The objective of this chapter is to discuss current progresses in research and development of the molecular mechanisms regulating appressorium morphogenesis and host-tissue penetration that is exemplified in a model of infection of rice plant by *Magnaporthe oryzae* causing rice blast disease. The scope of this review includes (but is not limited to) appressorium induction and differentiation, appressorium maturation and function. In addition, *M. oryzae* is an etiological factor of severe rice blast disease affecting the production of grain (e.g., rice and wheat) globally.

There are several recent reports on progress in the field of fungal development and infection [2–4]. We suggest that comprehensive understanding of the *M. oryzae* infection mechanisms may have a humanitarian impact by promoting development and implementation of effective strategies of the disease management, and thus by improving food supply.

Tissue Barriers in Disease, Injury and Regeneration
ISBN: 978-0-12-818561-2
https://doi.org/10.1016/B978-0-12-818561-2.00002-3

2. Appressorium induction and differentiation

Conidia, acting as main infection source, play important roles in initializing the infection cycle. In *M. oryzae*, upon conidial attachment on rice leaves, conidia germinate soon within one or two hours under humid condition and proper temperature. The tip of germ tube swells into a hook-shape structure and later differentiate into forming appressorium, a specialized infection structure [5].

2.1 Surficial recognition

Several membrane receptors on *M. oryzae* germ tube recognize surficial properties of touching face, including surface hardness, hydrophobicity, and plant wax layer. In *M. oryzae*, a putative sensor protein MoMsb2 is responsible to sense surface hydrophobicity and cutin monomer, while another sensor protein MoSho1 has a potential role in recognizing primary alcohols, a component of rice leaves wax [6]. Further analysis reveals that the MoMsb2 protein interacts with Ras2, a GTPase, and functions upstream of the PMK1 MAPK pathway, signaling appressorium formation and penetration relatively independently through extracellular and cytoplasmic regions of MoMsb2 [7]. Pth11, a G protein-coupled receptor, is localized on cell membrane and governs appressorium formation when sensing inductive substrates cues such as cutin monomers and surface hydrophobicity. Disruption of functions in Pth11 caused failure in appressorium maturation although the Δ*pth11* mutant could still form germ tube hooks [8]. Recently, by structural function analysis, Kou, *et al.* reported that a CFEM domain at nitrogen terminus of Pth11 is required for appressorium formation and virulence in *M. oryzae* [9]. In addition, they also found another roles of Pth11 in regulating redox homeostasis during appressorium formation. A *PTH11* like G-protein coupled receptor (GPCR), WISH, was recently found to be required for surface sensing, resulting in defects in appressorium morphogenesis and pathogenesis [10]. Thus, appressorium differentiation might be orchestrated by many kinds of membrane-localized receptors.

2.2 G-protein/ Protein kinase C signaling pathway

Appressorium morphogenesis is regulated by G protein/PKA signaling pathway. G protein complex is a heterotrimer, composed of α, β, γ subunits, activated by upstream receptors and transducing signals from extracellular cues to intracellular receptors. There are three G protein α subunits

(MagA, MagB, and MagC), two G protein β subunits (Mgb1 and Mgb2), and one G protein γ subunit in *M. oryzae* [11]. Previous characterization of three G protein α subunits revealed that deletion of MagB impaired appressorium formation [12]. MoMgb1 is also required for producing appressoria and pathogenicity in *M. oryzae* [13]. Regulators of G protein signaling play negative roles in regulating activities of G protein. Rgs1 regulates appressorium formation and other development processes by binding all three G protein α subunits [11]. By systematic analysis of other 7 RGS-like proteins, MoRgs2, MoRgs3, MoRgs4, MoRgs6, and MoRgs7 are also involved in controlling appresorium formation [14]. Further findings reveal that MoRgs7 regulates cAMP signaling by a scaffold protein Mip11, which activates G protein MagA and inhibits activity of phosphodiesterase PdeH [15]. One important downstream effector of G protein complex is adenylate cyclase, which catalyzes the synthesis of cyclic AMP activating protein kinase A. Deletion of Mac1 in *M. oryzae* caused serious appressorium formation defects, which could be restored by adding exogenous cAMP [16]. The intracellular cAMP amount could also be regulated by phosphodiesterases thereby affecting activities of PKA signaling pathway. Similar to Rgs proteins, MoPdeL and MoPdeH mediated intracellular cAMP levels to further control appressorium formation [17]. Interestingly, MoImd4, an inosine-5′-monophosphate dehydrogenase, was explored to interact with MoPdeH. The crosstalk between MoImd4 and MoPdeH enhances phosphodiesterase enzyme activity, and further affected cAMP/PKA signaling pathway [18].

2.3 Protein kinase C signaling pathway

Protein kinase C (PKC) is conserved kinase in all eukaryotes and involved in activation and regulation of signal transduction pathways associated with growth, development, and cell death. Recently, several articles report that Mps1-MAPK signaling pathway also regulates appressorium development in filamentous fungi. Mps1, a core component of the cell wall integrity pathway and homolog of yeast Mpk1, play necessary roles in appressoria to repolarise and breach the host cuticle in rice blast fungus *M. oryzae* [19]. In *Colletotrichum lagenarium* and *Colletotrichum heterostrophus*, it also promotes the early phase of appressorium differentiation [20,21]. In addition, some transcriptional factors regulating appressorium formation in *M. oryzae*. The two downstream targets of Mps1-MAPK cascade are Mig1 and Swi6 essential in regulating appressorium morphogenesis and cell wall integrity [22]. Although OSM1-MAPK1 pathway plays a dispensable role in glycerol accumulation, appressorium turgor generation and pathogenicity, the downstream protein MoMsn2 plays an essential role in the appressorium development and enhance the pathogenicity of *M. oryzae,* suggesting that MoMsn2 regulates some downstream genes in stress response [23].

2.4 Glucose-ABL1-TOR signaling

Recently, Wilson *et al.* showed that glucose-ABL1-TOR signaling mediates appressorial cell differentiation by modulating cell cycle. The TOR signaling as a negative regulator of appressorium development that blocks cAMP/PKA signaling. When the activity of TOR was inhibited in the Δ*abl1* mutants, or the wild-type strains was treated with exogenous glucose, appressorium formation can be found by arresting cell at G1/G0 via inducing cell cycle delay at G2/M. These results reveal an underlying mechanism by which glucose regulates cell cycle by a glucose-ABL1-TOR signaling pathway [24].

3. Appressorium maturation and function

3.1 Autophagy

Autophagy is an intracellular physiological process involved in turnover of obsolete proteins or damaged organelles. Along with 2016 Nobel Prize in Physiology or Medicine awarded to autophagy research field, more researchers focus on depicting roles of autophagy in fungal development and pathogenicity. Autophagy has been reported to be required for virulence in many pathogenic fungi [25]. In *M. oryzae*, conidial autophagy occurs before appressorium maturation to promote turgor generation in appressoria [26]. Systematic analysis of 22 autophagy-related genes confirms that 16 components from nonselective macroautophagy are necessary for appressorium formation and virulence, while those involved in selective autophagy are not important for appressorium functions [27]. MoAtg14, as a new core component, was identified to be conserved in autophagy function but be divergent in amino acid sequence, indicated an evolution from yeast to mammalian cells [28]. Recent studies in *M. oryzae* have also evidenced that autophagy is tightly regulated by a great deal of factors in response to extrinsic environmental conditions. For example, the retromer complex functioning in intracellular cargo trafficking participates in autophagosome formation [29]. An epigenetic factor MoSnt2 was found to mediate acylation of histone 3 to influence transcription of autophagy-related proteins MoAtg6, 15, 16, and 22 [30]. In addition, MoAtg3 and MoAtg9 are also directly modified by an acetyltransferase MoHat1-mediated posttranslational modification to regulate appressorium formation and virulence [31]. In future, specific regulation mechanisms involving initiation and termination of autophagy will be depicted from different point of view.

3.2 Melanin, lipid, and glycogen metabolism

During appressorium maturation, some melanin synthesis-related genes are expressed and catalyze formation of a thick differentiated melanin layer on the inner side of the appressorium cell wall, which is required to prevent efflux of glycerol from the rapidly expanding appressorium and also to provide structural rigidity and resilience to break down plant cuticle [32]. Maturation of the appressorium in *M. oryzae* undergoes rapid accumulation of glycerol to form enough turgor pressure [33]. When the conidia germinated, storage carbohydrates such as glycogen and trehalose are quickly degraded and mobilized to the appressorium and promote appressorium development [34].

3.3 Septin-mediated polarization

Septin is known as a kind of GTPase which takes part in several physiological processes, such as cytokinesis in hyphae, intracellular transport, reorientation, and apoptosis. Usually, septins regulate these processes by acting as a scaffold to recruit proteins or diffusion barriers to localize and concentrate related macromolecular substances. In *M. oryzae*, septins separate the germ tube and appressorium by localizing the myosin to the septum. Recently, it has been shown that the appressorium-mediated penetration of *M. oryzae* is dependent on septins-mediated polarization. A heteromeric septin ring, polymerized by Sep3, Sep4, Sep5, and Sep6, facilitates the toroidal F-actin network organized at the appressorium pore. Then, a penetration peg emerges from the appressorium pore under a huge hydrostatic pressure caused by accumulation of glycerol in appressorium [35]. The exocyst is one of the multiple tethering complex, mediating the tethering of post Golgi secretory vesicles to the plasma membrane and promotes the assembly of the SNARE complex for membrane fusion, which is strongly associated with polarized, hyphal growth [36]. In *M. oryzae*, using immunoprecipitation and liquid chromatography-tandem mass spectrometry to analyze the components of exocyst, it is found that four core septin GTPases has been detected. They finally confirmed that septin-dependent assembly of exocyst is necessary for polarized growth of fungi and plant infection by *M. oryzae*.

Recently, a new turgor sensor, Sln1, which is a kind of histidine–aspartate kinase has been discovered. The researchers find it necessary in sensing the turgor pressure in appressorium. When the pressure reaches to a threshold, this turgor sensor will regulate the emergence of penetration peg. An isogenic mutant of Sln1 gives an excess intracellular appressorium turgor and results in producing nonfunctional hyper-melanized appressorium. It is notable that one of the reasons causing the mutant defective in generating penetration peg is the failure in organization of the septins. And then, these influence the repolarization of appressorium which is required for leaf infection [37].

Actually, the function of septins is not only in polarization and plant infection but also in the initial phases of appressorium development, cytokinesis, and so on. More and more location evidences of septins have been revealed. Therefore, septins play critical roles in many physiological processes especially in the penetration. The critical functions of septins prompt us to carry out more significant septins researches to deal with the harm of rice blast.

4. Appressorium penetration and invasive growth

4.1 The MAPK PMK1 signaling pathway

The cAMP signaling pathway is mainly involved in surficial recognition activating germination and appressorium differentiation. In comparison, a conserved MAPK signaling pathway mediated by an Mst11-Mst7-Pmk1 cascade regulates appressorium maturation and penetration. Previous studies revealed that Mst50 acts as a scaffold linking MAPKKK Mst11 and MAPKK Mst7 [38]. Ras2 could interact with Mst50 to regulate appressorium formation. Further, the mechanism was confirmed that activated Ras2 results in activation of Pmk1 pathway and cAMP/PKA pathway to form abnormal appressorium. Now, Mst50 is confirmed to be involved in multiple MAP kinase signaling pathways [39]. In addition, Mst11 also binds with Ras2 and shows stronger binding ability with activated Ras2 to partly suppress cAMP signaling pathway preventing appressorium redifferentiation [40]. The MAPK signaling pathway is also mediated by the PKA activity. A transcriptional factor Sfl1 downstream from Pmk1 was found to be suppressed by PKA-mediated phosphorylation [41]. Recent research findings shed light on novel functions of the Pmk1 mitogen-activated protein kinase. Sakulkoo *et al.* found that Pmk1 regulates invasive hyphal growth from cell to neighbor cells. In addition, effectors responding to inhibit host immunity are also controlled by Pmk1 [42].

4.2 The MAPK Mps1 signaling pathway

The MAPK signaling cascade Mck1-Mkk1-Mps1 is involved in maintenance of cell wall integrity, appressorium penetration, and invasive growth. By expression of a plant MAPK inhibitor HopAI, an effector from *Pseudomonas syringae*, in *M. oryzae*, MAPK Mps1 is suppressed and leads to defects in invasive hyphal growth, revealing a critical role of Mps1 in host cell spreading [43]. A serine and threonine protein phosphatase Ppe1 was found to interact with Mkk1 to regulate Mps1 phosporylation in response to nitrogen starvation [44]. Crosslink also exists between cAMP/PKA signaling pathway and MAPK signaling pathways. Deletion of a phosophodiesterase PdeH causes dephosphorylation of Mps1 and phosphorylation of Pmk1, indicating cAMP/PKA signaling pathway positively regulated Pmk1 and negatively regulated Mps1 [45]. Recently, Yin *et al.* revealed that autophagy-related gene Atg1, a kinase, could directly phosphorylate MAPKK Mkk1 to respond to endoplasmic reticulum stress during plant infection [46].

4.3 Effectors

To invade plant host, pathogenic fungi evolve different structures by which deliver effectors into host cells to suppress plant immunity and colonize host cells. Effectors are small secreted proteins less than 200 amino acids, highly upregulated expression during plant colonization. Based on localization during invasion, effectors are mainly divided into two types depending on two different secretion systems delivered into host system, including apoplastic effectors and cytoplasmic effectors [47]. Many effectors have been identified and summarized in a previous review [2]. Some of them have been depicted in detail that they involve in infection, such as AvrPiz-t and AvrPiiA. However, most of them have been found to be required for virulence of rice blast fungus, the specific mechanism by which they operate remain obscure and are worth investigating in the future. By transient expression of putative secreted proteins in *Nicotiana benthamiana*, a total of 13 cell death-inducing factors were identified and named MoCDIP1-13 [48]. These effectors may be critical for early biotrophic stage of infection. A recent report has identified another effector MoSVP which is a novel virulence-related factor by analyzing transcriptome profile of the initial stage of infection [49].

5. Conclusion

Appressorium development and differentiation is an extremely complicated process that is coordinated by a network of proteins and a variety of signaling pathways to invade the host plant successfully. In addition, many proteins from that network can play dual roles; and there is crosstalk among the signaling pathways. In general, comprehension of the interactive networks of proteins and signaling pathways driving appressorial differentiation and penetration can be helpful for R&D of new strategies to control rice blast disease.

Acknowledgments

This study was supported by grant from National Science and Technology Major Project (2018ZX08001-03B).

References

[1] Talbot NJ. Appressoria. Curr Biol (CB) 2019;29:R144–6.
[2] Yan X, Talbot NJ. Investigating the cell biology of plant infection by the rice blast fungus *Magnaporthe oryzae*. Curr Opin Microbiol 2016;34:147–53.
[3] Jiang C, Zhang X, Liu H, Xu JR. Mitogen-activated protein kinase signaling in plant pathogenic fungi. PLoS Pathog 2018;14. e1006875.
[4] Deng YZ, Naqvi NI. Metabolic basis of pathogenesis and host adaptation in rice blast. Annu Rev Microbiol 2019;73:601–19.
[5] Talbot NJ. On the trail of a cereal killer: Exploring the biology of *Magnaporthe grisea*. Ann Rev Microbiol 2003;57:177–202.
[6] Liu W, Zhou X, Li G, Li L, Kong L, Wang C, et al. Multiple plant surface signals are sensed by different mechanisms in the rice blast fungus for appressorium formation. PLoS Pathog 2011;7:e1001261.
[7] Wang GH, Li GT, Zhang SJ, Jiang C, Qin J, Xu JR. Activation of the signalling mucin MoMsb2 and its functional relationship with Cbp1 in *Magnaporthe oryzae*. Environ Microbiol 2015;17:2969–81.
[8] DeZwaan TM, Carroll AM, Valent B, Sweigard JA. *Magnaporthe grisea* Pth11p is a novel plasma membrane protein that mediates appressorium differentiation in response to inductive substrate cues. Plant Cell 1999;11:2013–30.
[9] Kou YJ, Tan YH, Ramanujam R, Naqvi NI. Structure-function analyses of the Pth11 receptor reveal an important role for CFEM motif and redox regulation in rice blast. New Phytol 2017;214:330–42.
[10] Sabnam N, Barman SR. WISH, a novel CFEM GPCR is indispensable for surface sensing, asexual and pathogenic differentiation in rice blast fungus. Fungal Genet Biol 2017;105:37–51.
[11] Liu H, Suresh A, Willard FS, Siderovski DP, Lu S, Naqvi NI. Rgs1 regulates multiple Galpha subunits in *Magnaporthe* pathogenesis, asexual growth and thigmotropism. EMBO J 2007;26:690–700.
[12] Liu S, Dean RA. G protein alpha subunit genes control growth, development, and pathogenicity of *Magnaporthe grisea*. Mol Plant Microbe Interact (MPMI) 1997;10: 1075–86.

[13] Nishimura M, Park G, Xu JR. The G-beta subunit MGB1 is involved in regulating multiple steps of infection-related morphogenesis in *Magnaporthe grisea*. Mol Microbiol 2003;50:231–43.

[14] Zhang H, Tang W, Liu K, Huang Q, Zhang X, Yan X, et al. Eight RGS and RGS-like proteins orchestrate growth, differentiation, and pathogenicity of *Magnaporthe oryzae*. PLoS Pathog 2011;7. e1002450.

[15] Yin Z, Zhang X, Wang J, Yang L, Feng W, Chen C, et al. MoMip11, a MoRgs7-interacting protein, functions as a scaffolding protein to regulate cAMP signaling and pathogenicity in the rice blast fungus *Magnaporthe oryzae*. Environ Microbiol 2018; 20:3168–85.

[16] Choi W, Dean RA. The adenylate cyclase gene MAC1 of *Magnaporthe grisea* controls appressorium formation and other aspects of growth and development. Plant Cell 1997;9:1973–83.

[17] Zhang H, Liu K, Zhang X, Tang W, Wang J, Guo M, et al. Two phosphodiesterase genes, PDEL and PDEH, regulate development and pathogenicity by modulating intracellular cyclic AMP levels in *Magnaporthe oryzae*. PLoS One 2011;6:e17241.

[18] Yang L, Ru Y, Cai X, Yin Z, Liu X, Xiao Y, et al. MoImd4 mediates crosstalk between MoPdeH-cAMP signalling and purine metabolism to govern growth and pathogenicity in *Magnaporthe oryzae*. Mol Plant Pathol 2019;20:500–18.

[19] Penn TJ, Wood ME, Soanes DM, Csukai M, Corran AJ, Talbot NJ. Protein kinase C is essential for viability of the rice blast fungus *Magnaporthe oryzae*. Mol Microbiol 2015; 98:403–19.

[20] Kojima K, Kikuchi T, Takano Y, Oshiro E, Okuno T. The mitogen-activated protein kinase gene MAF1 is essential for the early differentiation phase of appressorium formation in *Colletotrichum lagenarium*. Mol Plant Microbe Interact 2002;15:1268–76.

[21] Eliahu N, Igbaria A, Rose MS, Horwitz BA, Lev S. Melanin biosynthesis in the maize pathogen *Cochliobolus heterostrophus* depends on two mitogen-activated protein kinases, Chk1 and Mps1, and the transcription factor Cmr1. Eukaryot Cell 2007;6:421–9.

[22] Qi Z, Wang Q, Dou X, Wang W, Zhao Q, Lv R, et al. MoSwi6, an APSES family transcription factor, interacts with MoMps1 and is required for hyphal and conidial morphogenesis, appressorial function and pathogenicity of *Magnaporthe oryzae*. Mol Plant Pathol 2012;13:677–89.

[23] Zhang H, Zhao Q, Guo X, Guo M, Qi Z, Tang W, et al. Pleiotropic function of the putative zinc-finger protein MoMsn2 in *Magnaporthe oryzae*. Mol Plant Microbe Interact 2014;27:446–60.

[24] Marroquin-Guzman M, Sun G, Wilson RA. Glucose-ABL1-TOR signaling modulates cell cycle tuning to control terminal appressorial cell differentiation. PLoS Genet 2017;13. e1006557.

[25] Liu XH, Gao HM, Xu F, Lu JP, Devenish RJ, Lin FC. Autophagy vitalizes the pathogenicity of pathogenic fungi. Autophagy 2012;8:1415–25.

[26] Veneault-Fourrey C, Barooah M, Egan M, Wakley G, Talbot NJ. Autophagic fungal cell death is necessary for infection by the rice blast fungus. Science 2006;312:580–3.

[27] Kershaw MJ, Talbot NJ. Genome-wide functional analysis reveals that infection-associated fungal autophagy is necessary for rice blast disease. Proc Natl Acad Sci U S A 2009;106:15967–72.

[28] Liu XH, Zhao YH, Zhu XM, Zeng XQ, Huang LY, Dong B, et al. Autophagy-related protein MoAtg14 is involved in differentiation, development and pathogenicity in the rice blast fungus *Magnaporthe oryzae*. Sci Rep 2017;7:40018.

[29] Zheng W, Zhou J, He Y, Xie Q, Chen A, Zheng H, et al. Retromer is essential for autophagy-dependent plant infection by the rice blast fungus. PLoS Genet 2015;11. e1005704.

[30] He M, Xu Y, Chen J, Luo Y, Lv Y, Su J, et al. MoSnt2-dependent deacetylation of histone H3 mediates MoTor-dependent autophagy and plant infection by the rice blast fungus *Magnaporthe oryzae*. Autophagy 2018;14:1543–61.

[31] Yin Z, Chen C, Yang J, Feng W, Liu X, Zuo R, et al. Histone acetyltransferase MoHat1 acetylates autophagy-related proteins MoAtg3 and MoAtg9 to orchestrate functional appressorium formation and pathogenicity in *Magnaporthe oryzae*. Autophagy 2019;15:1234–57.

[32] Foster AJ, Ryder LS, Kershaw MJ, Talbot NJ. The role of glycerol in the pathogenic lifestyle of the rice blast fungus *Magnaporthe oryzae*. Environ Microbiol 2017;19:1008–16.

[33] Talbot NJ. On the trail of a cereal killer: Exploring the biology of *Magnaporthe grisea*. Annu Rev Microbiol 2003;57:177–202.

[34] Dixon KP, Xu JR, Smirnoff N, Talbot NJ. Independent signaling pathways regulate cellular turgor during hyperosmotic stress and appressorium-mediated plant infection by *Magnaporthe grisea*. Plant Cell 1999;11:2045–58.

[35] Dagdas YF, Yoshino K, Dagdas G, Ryder LS, Bielska E, Steinberg G, Talbot NJ. Septin-mediated plant cell invasion by the rice blast fungus, *Magnaporthe oryzae*. Science 2012;336:1590–5.

[36] Gupta YK, Dagdas YF, Martinez-Rocha AL, Kershaw MJ, Littlejohn GR, Ryder LS, et al. Septin-Dependent Assembly of the Exocyst Is Essential for Plant Infection by *Magnaporthe oryzae*. Plant Cell 2015;27:3277–89.

[37] Ryder LS, Dagdas YF, Kershaw MJ, Venkataraman C, Madzvamuse A, Yan X, et al. A sensor kinase controls turgor-driven plant infection by the rice blast fungus. Nature 2019;574:423–7.

[38] Park G, Xue C, Zhao X, Kim Y, Orbach M, Xu JR. Multiple upstream signals converge on the adaptor protein Mst50 in *Magnaporthe grisea*. Plant Cell 2006;18:2822–35.

[39] Li G, Zhang X, Tian H, Choi YE, Tao WA, Xu JR. MST50 is involved in multiple MAP kinase signaling pathways in *Magnaporthe oryzae*. Environ Microbiol 2017;19:1959–74.

[40] Qi L, Kim Y, Jiang C, Li Y, Peng Y, Xu JR. Activation of Mst11 and Feedback Inhibition of Germ Tube Growth in *Magnaporthe oryzae*. Mol Plant Microbe Interact 2015;28:881–91.

[41] Li Y, Zhang X, Hu S, Liu H, Xu JR. PKA activity is essential for relieving the suppression of hyphal growth and appressorium formation by MoSfl1 in *Magnaporthe oryzae*. PLoS Genet 2017;13:e1006954.

[42] Sakulkoo W, Oses-Ruiz M, Oliveira Garcia E, Soanes DM, Littlejohn GR, Hacker C, et al. A single fungal MAP kinase controls plant cell-to-cell invasion by the rice blast fungus. Science 2018;359:1399–403.

[43] Zhang X, Liu W, Li Y, Li G, Xu JR. Expression of HopAI interferes with MAP kinase signalling in *Magnaporthe oryzae*. Environ Microbiol 2017;19:4190–204.

[44] Qian B, Liu X, Jia J, Cai Y, Chen C, Zhang H, et al. MoPpe1 partners with MoSap1 to mediate TOR and cell wall integrity signalling in growth and pathogenicity of the rice blast fungus *Magnaporthe oryzae*. Environ Microbiol 2018;20:3964–79.

[45] Yang LN, Yin Z, Zhang X, Feng W, Xiao Y, Zhang H, et al. New findings on phosphodiesterases, MoPdeH and MoPdeL, in *Magnaporthe oryzae* revealed by structural analysis. Mol Plant Pathol 2018;19:1061–74.

[46] Yin Z, Feng W, Chen C, Xu J, Li Y, Yang L, et al. Shedding light on autophagy coordinating with cell wall integrity signaling to govern pathogenicity of *Magnaporthe oryzae*. Autophagy 2019:1–17.

[47] Giraldo MC, Dagdas YF, Gupta YK, Mentlak TA, Yi M, Martinez-Rocha AL, et al. Two distinct secretion systems facilitate tissue invasion by the rice blast fungus *Magnaporthe oryzae*. Nat Commun 2013;4:1996.
[48] Guo X, Zhong D, Xie W, He Y, Zheng Y, Lin Y, et al. Functional Identification of Novel Cell Death-inducing Effector Proteins from *Magnaporthe oryzae*. Rice (N Y) 2019;12:59.
[49] Shimizu M, Nakano Y, Hirabuchi A, Yoshino K, Kobayashi M, Yamamoto K, et al. RNA-Seq of in planta-expressed *Magnaporthe oryzae* genes identifies MoSVP as a highly expressed gene required for pathogenicity at the initial stage of infection. Mol Plant Pathol 2019;20:1682.

CHAPTER SIX

Mesenchymal stem cells and exosomes in tissue regeneration and remodeling: characterization and therapy

Juliann G. Kiang[1,2,3]
[1]Scientific Research Department, Armed Forces Radiobiology Research Institute, Uniformed Services University of the Health Sciences, Bethesda, MD, United States
[2]Department of Pharmacology and Molecular Therapeutics, Uniformed Services University of the Health Sciences, Bethesda, MD, United States
[3]Department of Medicine, Uniformed Services University of the Health Sciences, Bethesda, MD, United States

1. Introduction

A growing number of in vivo and in vitro data provides new insights on great potentials of adult mesenchymal stem cells (MSCs) in regenerative medicine and their application for countermeasures against trauma and tissue injury. MSCs are ubiquitous component of tissue stroma, which constitutes the intrinsic structural barriers onto parenchyma of tissues and organs. The recent paradigm on biology of MSCs suggests that the cells can also express diverse mechanisms promoting recovery and remodeling of the stroma and parenchyma in response to stress and damage.

MSCs have been immensely investigated by using them for various applications, including tissue modification, repair, and regeneration on cell-based therapies [1,2,3]. These cells can be isolated from umbilical cord blood [4,5,6,7], placenta [7,8,9,10], bone marrow [2,7,11,12], or fat tissues [5,7,13,14,15,16,17,18]. Other tissues such as liver and chorionic tissues were used to extract MSCs as well [7]. Fat tissues are getting more prominent for MSC collection because of their easy access.

Characteristics of MSCs after collection and culturing is first to be performed, including attachment to cell culture dishes, fibroblast-like morphology, high-proliferation rate, colony formation, and capacities to differentiate into different mesenchymal lineages. They have been applied clinically to control autoimmune and graft-versus-host diseases [19,20,21,22],

Tissue Barriers in Disease, Injury and Regeneration
ISBN: 978-0-12-818561-2
https://doi.org/10.1016/B978-0-12-818561-2.00005-9

myocardial infarction [23,24], cartilage defects, osteoarthritis [25], and others [2]. In preclinical studies, MSCs have been shown to mitigate myocardiac infarction [26,27], promote healing in irradiated murine skin wound [11,28], protect against radiation-induced liver injury [29], improve survival in irradiated mice [2,11,30], alleviate the gastrointestinal syndrome in mice [31], repair the intestinal mucosal barrier in irradiated mice [32], and decrease cartilage defects and osteoarthritis [25], lung injury [33], polycystic kidney disease [34], acute graft-versus-host disease [35], and cognitive dysfunction [36,37]. Recently, there are reports indicating that MSC-derived extracellular vesicle transplantation repairs the irradiated brain [37,38,39].

MSC modification can reinforce MSC capabilities. It is reported that superoxide dismutase gene-transfected MSCs improved survival in irradiated mice [40]. In contrast to this report, other reports showed MSCs alone did not improve survival in irradiated mice [2,11]. MSCs overexpressing angiotensin-converting enzyme 2 (ACE 2) result in a further improvement after lung injury [33].

Recent understanding in the cellular and molecular signaling activations on adult MSCs have provided new insights into their potential clinical applications, particularly for tissue repair and regeneration. This chapter focuses on these advances on MSC characterization and therapeutic uses.

2. Mesenchymal stem cell characterization

MSCs have been present in nearly all postnatal tissues or organs, including umbilical cord blood [4–6], placenta [8,9,10], fat tissue [5,13,14,15,16,17,18], and bone marrow [11,12], among others. MSCs represent a progenitor population with multiple differentiation potentials [41,42,43,44,45,46,47,48]. With the advantage of autologous transplantation that avoids the immune rejection and ethical concerns, MSCs have great application prospect in personalized treatment of diseases [49,50,51]. Identification and purification of MSCs are important for MSCs efficacy. Cell membrane surface of MSCs displays specific markers. Therefore, these markers can be used to verify the identity of MSCs, while cell surface markers present in other type of cells such as hematopoietic progenitor stem cells are used as negative marker controls.

2.1 Mesenchymal stem cell negative markers

Friedenstein and colleagues in 1970 reported that MSCs have been identified as colony-forming unit-fibroblasts (CFU-Fs) [52]. In 1999, Pittenger and colleagues described in detail the tri-lineage potential of MSCs [53]. Today, our understanding of these cells has greatly advanced. MSCs are multipotent, adherent, and can be isolated from many adult tissue types. To ensure the isolated cells are MSCs, there is a consensus [3] that MSCs do not express glycophorin-A (an erythroid linerage marker), CD11b (an immune cell marker), CD31 (an endothelial and hematopoietic cell maker), CD34 (a primitive hematopoietic stem cell marker), CD45 (a marker of all hematopoietic cells), and CD117 (a hematopoietic stem/progenitor cell marker).

While CD34 surface marker is certainly negative in human MSCs, CD11b, CD45, and CD117 are certain MSC negative markers in both human MSCs and murine MSCs. In contrast, CD34 has also been found to be positive in murine MSCs [3]. For murine MSCs verification, multiple negative and positive surface markers are used because of this uncertainty of markers in murine MSCs.

2.2 Mesenchymal stem cell positive markers

Known MSC positive markers are used to ensure that isolated cells are MSCs. There are Stro-1, CD10, CD13, CD29, CD44, CD73, CD90/Thy-1, CD105, CD106 (i.e., VCAM-1), CD271/NGFR, Flk-1/CD309, and Sca-1. Among them, Stro-1 is the best-known MSC marker by far because Stro-1 negative cells do not form colonies [54]. Nonetheless, its surface expression in MSCs is gradually lost during culture expansion [50], through yet unidentified mechanism(s). Whether the loss of stro-1 marker resulting in the loss of colony capability is not clear. However, identification of MSCs always includes Stro-1 in conjunction with other MSC positive and negative marker proteins.

Ref. [40] identified MSCs with positive expression of CD13, CD29, CD44, CD105, and Sca-1 and negative expression of Thy-1.2, c-kit, CD11b, CD19, CD31, CD34, CD45, CD73, and CD135, while Ref. [29] identified MSCs with positive expression of CD105 and CD73 and negative expression of CD45. Ref. [31] identified MSCs with positive expression of CD29 and CD105 and negative expression of CD11b and CD133. Ref. [55] identified MSCs with positive marker proteins CD73, CD90, CD105, and CD146, and negative marker proteins CD11b, CD14, CD34, and CD45. Ref. [34] characterized MSCs with positive CD90 and CD29 expressions and negative CD45 and CD11b/c expression. Ref. [35] characterized MSCs with positive Sca-1, CD29, CD44, CD90.2, and Flk-1 expression and negative CD117 and CD34 expression. Ref. [11] identified murine bone marrow MSCs with positive expression of Strol-1, Sca-1, CD44, and CD105 and negative expression of CD3 and CD34, while MSCs displayed colony formation. Ref. [56] defined MSCs with positive for CD105, CD73, and CD90 and negative for CD45, CD34, CD14 or CD11b, CD79α or CD19, and HLA-DR surface markers. Fig. 6.1 shows that MSCs derived from mouse bone marrow. The average size of each individual cell is 5.5 ± 0.3 μm (n = 19). A representative BMSC colony is present.

2.3 Mesenchymal stem cell self-renewal and maintenance

MSCs are capable of self-renewal without differentiation. These cells express the embryonic stem cell gene markers oct-4, sox-2, and rex-1 [57] that are involved in repressing differentiation genes [58]. In addition, the presence of leukemia inhibitory factor [59,60], fibroblast growth factors [61,62], and mammalian homologs of Drosophila wingless [63] is observed. Hepatocyte growth factor (HGF), epidermal growth factor (EGF), platelet-derived growth factor (PDGF), and/or cytokines support MSC stemness in an

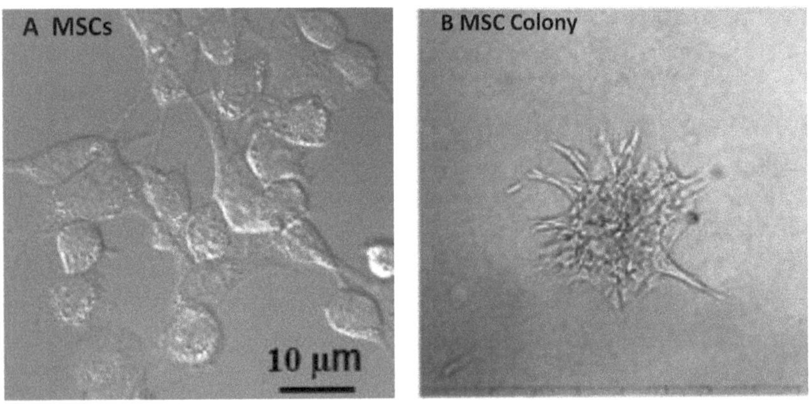

Figure 6.1 ***Mesenchymal stem cell (MSC) individual cell size and colony formation.*** MSCs derived from B6D2 F1 female mice (14–16 weeks old, Jackson Laboratories) were seeded in 4-well chambers at 4500 cells per well (Lab-Tek Chambered #1.0 Borosilicate Coverglass System, Fisher, Rochester, NY). On day 10, the 4-well chambers were rinsed with phosphate-buffered saline (PBS) before the addition of methanol. After sitting for 5 min, the chambers were again rinsed with PBS. The well cover was removed, and when dry, one drop of mounting medium (Vector Laboratories, Inc.; Burlingame, CA) was then added to each well and a cover slip was applied. (A) Representative MSCs. (B) Representative colony.

MSC niche [3]. Beta-catenin, an extracellular matrix protein for anchoring cells in place, is thought to be involved in Wnt regulation of MSC self-renewal [64].

The underlying molecular mechanisms of MSC differentiation still remain unclear, but some of factors for tissue repair and regeneration have been unfolded. MSC differentiation needs to be induced clinically by administration of transforming growth factor-beta (TGF-beta), bone morphogenetic protein (BMF), growth and differentiation factor [65], and Wnt ligands [66] for chondrogenesis, tenogenesis, and osteogenesis, peroxisome proliferator-activated receptor gamma (PPARgamma) for adipogenesis [67], and Notch 1 for myogenesis [68,69], respectively. To keep in mind, the differentiation signal must find its way to the MSC niche for initiation of differentiation. The detailed process requires further studies [3].

2.4 Mesenchymal stem cells proliferate in hypoxia faster than in normoxia

The proliferation rate of MSCs depends on the environment where these cells reside. Cells proliferated at an approximate rate of 0.4458×10^5 cells/h and 1.2067×10^5 cells/h under normoxia and hypoxia, respectively, when approximate 1×10^6 cells were seeded and cultured (Kiang, Ho, Smith, unpublished data). It was reported that it took 32–35 h [11,70] to double MSC numbers under normoxia. The results suggest that MSCs proliferate faster in low-oxygen tension than in normal oxygen tension, a phenomenon similar to cancer stem cells [71], yet the underlying mechanisms leading to this accelerated cell proliferation is not identified.

2.5 Mesenchymal stem cells isolated from different tissues are not equal

It is evident that MSCs can be obtained from different tissues in the body. In fact, MSCs isolated from different tissues are not functionally the same. Ref. [7] reported that the expression of typical MSC-associated gene THY1 and surface markers CD90 and CD73 were mostly similar between MSCs from different donor sites, while CD105 levels were various from 4.1% amniotic tissues-derived MSCs to 57.9% on liver-derived MSCs. Moreover, CD34 expression was low among six tissues, but CD45 expression was high in MSCs derived from chorionic tissue and liver.

The potential of MSCs from different donor sites to differentiate toward adipogenic, chondrogenic, and osteogenic lineages are compared. These authors report that MSCs from adipose tissue, bone marrow, and chorion differentiated to adipogenic, chondrogenic, and osteogenic lineages; MSCs from liver differentiated to chondrogenic and osteogenic lineages; MSCs from umbilical cord differentiated little at all.

Secretion of growth factors is different as well. For example, net HGF concentrations were drastically high in -chorionic tissue and liver tissue; net TIMP2 and net bFGF concentrations were immensely downregulated, while these growth factors were little or moderately changed in MSCs from other four donor sites. Therefore, the optimal sourcing for therapeutic transplantation should be taken into consideration because the encasement of MSCs into semipermeable membranes possibly provides a physical immune barrier so as to preventing cell fusion.

2.6 Mesenchymal stem cells kill bacterial by autophagy

MSCs usually exhibit appreciably high amounts of constitutively expressed heat shock protein 70 kDa (HSP70) and NF-keppaB-p65, and a detectable amount of NAD^+-dependent deacetylase sirtuin-3 (Sirt3). In our laboratory, significant rises in HSP70, NF-keppaB-p65, Sirt3, and matrix metalloproteinase-3 (MMP3) were observed, when MSCs were exposed to ^{60}Co γ radiation at 12 Gy but not 8 Gy. Sirt3 is a mitochondrial stress-response protein. Rises in Sirt3 expression indicate that radiation induces stress to mitochondria. Caspase-3, a marker for caspase-dependent apoptosis, was not detected in irradiated MSCs, indicating that no apoptosis occurs in MSCs after irradiation. Radiation induced significant increases in light chain 3 (LC3) expression, a marker of autophagy, detected by Western blotting and LC3-containing autophagy vacuoles displayed by immunofluorescent staining, suggesting presence of up-regulation of autophagy defense machinery [72].

Radiation is known to cause systemic bacterial infection [73,74,75]. The immune homeostasis and defense response to blood pathogens are mediated by the marrow-blood barrier in bone marrow, which is composed of endothelial, reticuloendothelial, and mesenchymal stromal cell lineages [76–79]. When MSCs were irradiated alone or followed by exposure to Gram-negative *E. coli* challenge (5×10^7 bacteria/mL), rises in lysosomal-associated membrane protein 1 (Lamp1), small ubiquitin-related modifier 1 (SUMO1), collagen III, MMP3, MMP13, and p62/SQSM1 were found 24 h after irradiation alone or combined with *E. coli* challenge. MSCs exerted extensive phagocytosis and inactivated bacteria in autolysosomes [80]. When MSCs were challenged with Gram-positive *S. epidermidis*

(5×10^7 bacteria/mL) for 3 h, the cells displayed remarkable resistance to the bacterial challenge and sustained confluence over the period of observation. Similar observations to that with *E. coli* challenge [81] or lipopolysaccharide (LPS) exposure [80] were found as well. We postulate that the above prosurvival pathways activated in MSCs in vitro could be adaptive responses employed by stromal cells under septic conditions. These results also suggest that MSCs can contribute to the innate defense response to radiation injury.

2.7 Mesenchymal stem cells exhibits mitochondrial remodeling

Radiation results in bacterial infection [75]. Radiation or bacterial challenge of MSCs results in alteration of the mitochondrial network. Electron transmission microscopy (ETM) and immunofluorescence microscopy exhibit that the normal mitochondrial network is a combination of round and elongated organelles containing discrete cristae at high densities. Mitochondrial fusion and fission occur when necessary.

Using ETM, MSCs with the bacterial challenge showed extensive mitochondrial swelling and cristae fragmentation 5 h postchallenge. The entire mitochondrial body almost became reticular by 24 h postchallenge. This structure rearrangement and fragmentation were triggered by increased expression of immunity-related GTPase family M (IRGM) and inducible nitric oxide synthase (iNOS). The bacterial challenge also induced dynamin-related protein 1 (Drp1, a marker of mitochondrial fission) translocation from cytosol to mitochondria, leading to activation of PTEN induced putative kinase 1- parkin RBR E3 ubiquitin protein ligase (PINK1-PARK2) to initiate mitophagy to degrade fragmented mitochondria [81]. Using Western blotting technique, the bacterial challenge resulted in significant increases in proteins of mitofusin-1 (Mfn1, a marker of mitochondrial fusion), PINK1, and PARK2 in MSCs 24 h after the bacterial challenge. The HSP70 basal level was not affected. Furthermore, no caspase-3 activation was detected in these cells [81]. These results, taken together, suggest that mitophagy but not caspase-dependent apoptosis in MSCs occurs after the bacterial challenge. It is warranted that caspase-independent apoptosis caused by molecular pathways involving with apoptosis-inducible factor or by senescence signals of protein 16 (p16) and beta-galactosidase (beta-gal) in MSCs after bacterial challenge shall be explored. MSCs exposed to ionizing irradiation alone also show mitochondrial fission and subsequent fusion as well as mitophagy [81].

2.8 Mesenchymal stem cells and signal transduction

In vivo, radiation deactivates AKT and activates JNK and p38 proteins in murine spleen [82] and deactivates AKT and activates p38 in murine small brain [82a] and small intestine [82b]. Moreover, radiation reduces ATP in small brain [82a]. Ref. [55] reported that NO activated c-Raf, JNK, p38, p53, and a nuclear factor E2-related factor (NRF2)-associated stress response in bone, which may have detrimental consequences for bone remodeling or bone regeneration. In vitro, our laboratory [Kiang et al., unpublished data] found that γ-radiation at 6 Gy decreased AKT activation and increased JNK activation in MSCs. Meanwhile, radiation reduced the cellular ATP and mitochondrial ATP. This reduction could be mediated through reduction of NRF1/2 and mitochondrial complexes 1-V [82a]. Therefore, one can postulate that priming MSCs to elevate prosurvival signaling molecules may enhance MSCs' multifunctionality as a therapy.

3. Mesenchymal stem cells and tissue or organ therapy

Organ dysfunction and failure has been a serious long-standing problem for patients with various kinds of diseases, aging degeneration/dysfunction, accidental trauma, or hereditary disorders. Organ transplantation is one of the choices for repair or clinically symptomatic amelioration. MSCs have been targeted to repair the problem using the laboratory animal models to develop methodologies and strategies. There are promising reports available.

3.1 Mesenchymal stem cells improve diabetes

Incidences of Type 1 and type 2 diabetes are increasing quickly because of the diet changes. Poorly controlled diabetes can lead to severe complications that can be life-threatening. Adult MSCs have been investigated for curing or alleviating this disease, hoping to get MSCs to become β-cells of Langerhans' island to secret insulin. Unfortunately, it is not successful. However, MSCs have been found to secrete many immunomodulatory and tissue regenerative factor [82c]. It is possible that the MSCs derived from bone marrow are not suitable for this purpose (See Section 2.5 above). MSCs derived from other donor sites should be explored and optimize the chance for reaching this proposed attempt. Any success in this case will bring hope to other diseases such as Parkinson disease, Multiple Sclerosis, Hodgeson Lymphoma, etc.

3.2 Mesenchymal stem cells improve acute lung injury

Acute lung injury is to have the diffuse damage of lung vascular endothelial cells and alveolar epithelial cells and an excessive inflammatory response in the lung [83]. It is evident that the renin-angiotensin system plays an essential role in causing this syndrome because of an increasing level of angiotensin II generated by the ACE. Ref. [33] reported that ACE2 gene transfected MSCs (5×10^5 cells/mouse) were injected into male wild-type and ACE2 knockout C57BL/6 mice that had been rendered to have acute lung injury. ACE2 gene-transfected MSCs transplantation 24 and 72 h later resulted in alleviation of the lung histopathology and presence of anti-inflammatory effects. The transplantation also reduced pulmonary vascular permeability, improved endothelial barrier integrity, and normalized lung eNOS expression. The report further indicates that ACE2 gene-transfected MSCs were recruited into the injured lung and enhanced local expression of ACE2 protein, thereby leading to decreases in angiotensin II levels accumulated in the lung.

The results in injured lung of Ref. [33] included decreases in neutrophil counts, IL-1β and IL-6, increases in IL-10, and decreases in pulmonary edema. These results are consistent with previous studies in that the MSCs were mainly taken up by the injured lung 4−7 days postinjection [84]. This improvement in acute lung injury in mice makes MSCs a potential therapy for treating this devastating clinical syndrome.

3.3 Mesenchymal stem cells improve renovascular function in kidney

Hereditary polycystic kidney disease is one of the most common lethal monogenic genetic diseases of man [85,86]. There is no acceptable treatment for this disease. Using a rodent experimental model [34], intrarenal infused MSCs at 2.5×10^5 cells in 250 μL of PBS into female Sprague−Dawley rats that were expressing polycystic kidney disease was performed. Four weeks later, MSCs preserved vascular density and glomeruli diameter, reduced fibrosis, preserved the expression of proangiogenic molecules. The benefit was observed up to four weeks after a single MSC infusion, suggesting that the cell-based therapy constitutes a novel approach in polycystic kidney disease, a lethal disease.

3.4 Differentiated Mesenchymal stem cells improve myocardiac performance

Cardiovascular disease, in particular, coronary heart disease, is the main disease type causing the majority of deaths. The damage to heart wall is irreversible. The current pharmacological treatments, surgical measures, and heart transplantation are in practice with known disadvantages. Ref. [87] reported that C57BL/6 male mice were injected s.c. with recombinant rat SCF, 200 μg/kg/day, and recombinant human G-CSF, 50 μg/kg/day once a day for five days before and three days after ligation of the coronary artery. Bone marrow cells (BMCs) were mobilized to form 15×10^6 new myocytes, decreased mortality by 68%, and reduced infarct size by 40%.

The differentiation of myocardiac cells from MSCs can be induced by biochemical, myocardial microenvironment, genetic modification, and traditional Chinese herbs [23], which provided a promising perspective to cell treatment on cardiac diseases. Ref. [24] reported the first case of the autologous MSCs transplantation for acute myocardiac infarction in clinical trials. This team transplanted 1×10^7 autologous MSCs into infracted artery by catheter. The transplantation was safe and preliminarily effective. The research team [27] injected 5×10^4 skeletal myoblasts by intramyocardial delivery system into sheep and months later the cardiac function was improved. Ref. [26] reported that the research team injected 5×10^6 GFP-expressing skeletal myoblasts from male Sprague–Dawley rats into female Sprague–Dawley rats with myocardiac infarction by either retrograde intracoronary or intramyocardial routes. These skeletal myoblasts improved cardiac performance and physical activity, associated with reduced cardiomyocyte-hypertrophy and fibrosis, further supporting the effectiveness of differentiated stem cells.

The meta-analysis from 2017 to 2019 [88] provided the cumulative efficacy and safety results based on randomized controlled trials. The cumulative efficacy analysis suggests that stem cell therapy is associated with a moderate improvement in left ventricular ejection fraction. The safety analysis indicates no increased risk of mortality in patients with advanced heart failure. However, to satisfy the clinical usage, it needs to ameliorate the conditions of induction and to further improve the differentiation efficiency. More investigation with a larger sample size is still desirable.

3.5 Mesenchymal stem cells improves radiation-induced damage in intestinal mucosal barrier

Radiation is known to induce intestinal damage [82b]. Ref. [32] reported that male CD2F1 mice were subjected to a dose of 8 Gy total body irradiation (Cs-137 irradiator, 1.35 Gy/min). Within 4 6 h after irradiation, these mice received an intravenous injection of 2×10^7 BMCs supplemented with 1×10^7 spleen cells. The transplantation accelerated recoverry of peripheral blood counts, enhanced the recovery of intestinal immune cell populations in jejunum mucosa, reduced intestinal permeability, reduced IL-1α increases, restored IL-6, IL-10, and IL-12 concentrations, and modulated the expression of Claudin-2 and -4 (tight junction proteins). Since whole BMCs were injected, whether MSCs were responsible for mitigation of intestinal mucosal barrier damage remains unclear and needs to be further studied. Whole bone marrow transplantation also showed the significant survival improvement in female B6D2F1 mice after irradiation [89].

Ref. [31] reported that male C57BL/6 mice were irradiated at 10 Gy (total body irradiation) or 16—20 Gy (abdominal irradiation) with a 320 KvP, Phillips MGC-40 orthovoltage irradiator (0.72 Gy/min). These mice were then intravenously injected with 2×10^6 MSCs per mouse at 24 and 72 h after irradiation. All MSC administered mice survived from 10 Gy or 16—20 Gy for more than 25 days, whereas irradiated mice administered with either the enriched myeloid fraction or the nonmyeloid factions failed to improve survival. MSCs induced crypt interstitial stem cell (ISC) regeneration, restitution of the ISC niche, and xylose absorption. R-Spondin1, KGF, PDGF, FGF2, and anti-inflammatory cytokines (G-CSF and Gm-CSF) were elevated in serum, while inflammatory cytokines (IL-6, IL-10, IL-12, and IL-17) declined.

Ref. [90] reported that BALB/c male and female mice were irradiated at 10 Gy (total body irradiation) with a ^{137}Cs-ray (o.662 MeV) irradiator at 1.05 Gy/min. These mice were then intravenously injected with 2×10^6 MSCs (derived from embryos) per mouse at 48 and 96 h after irradiation. Treatment with MSCs increased median survival time and significantly improved ileum morphology seven days postirradiation, suggesting regeneration of the tissue barrier or remodeling.

3.6 Mesenchymal stem cells repair radiation-induced liver injury

Radiation induces liver injury that can be detected by elevation of AST and ALT. Ref. [29] reported that NOD/SCID mice were irradiated with Cs-137 at 3.2 Gy (at 1.85 Gy/min) and then intravenously administered with 5×10^6 human MSCs in 0.1 mL 1X PBS. They indicated that MSC administration alone did not produce liver toxicity. MSC transplantation restored plasma urea, reduced plasma AST and ALT, and decreased the oxidative stress indicated by malondialdehyde formation.

They reported that stromal cell-derived factor 1 (SDF1) secreted by cells within injured tissues and its receptor C-X-C chemokine receptor type 4 (CXCR4) were necessary for the MSC migrating to damaged tissues. Livers of MSC administered mice displayed high levels of SDF1 and CXCR4 with reduction of mir-27b after irradiation. The latter is known to down-regulate SDF1. It took 15 days for MSCs to differentiate into the hepatocyte phenotype as indicated by measuring liver specific genes such as CK18, CK19, and AFP [29].

3.7 Mesenchymal stem cells accelerate the radiation-induced delay in wound healing

It is evident that radiation delays skin wound healing [28,91]. Ref. [28] reported that male Sprague—Dawley rats were exposed to 6 Gy of ^{60}Co gamma-ray (0.31 Gy/min) followed by a full-thickness excisional skin-wound (2% total body surface area). Then 1×10^7 recombinant adenovirus Adv-hPDGF-A/hBD2-GFP-infected MSCs (T-MSCs) or nontransfected MSCs (N-MSCs) were injected into the wound bed and margin of the excisional wound. These authors indicated that wounds in nonirradiated rats and irradiated rats took 17—18 days and 27—28 days, respectively, to heal. T-MSC administration and N-MSC administration were associated with a shorter healing time of 21 days and 24—25 days, respectively. MSCs promoted the deposition and remodeling of collagen in wounds. Significantly less bacterial colony formation was found in the cultured under-scar samples from the T-MSC administered wound bed. In our laboratory, when female B6D2F1 mice were exposed to 9.25 Gy of ^{60}Co gamma-ray (0.4 Gy/min) followed by a full-thickness excisional skin-wound (15% total body surface area). MSCs (3×10^6 cells) were intravenously injected. Their wounds were fully closed by day 21 after irradiation, whereas wounds in vehicle-treated irradiated mice were not fully healed yet at this time [11]. Our results are in agreement with observations reported by other laboratories [28].

The benefit of MSCs can be a crucial factor in large animals as in small animals. Ref. [92] reported that minipigs were locally irradiated at a dose of 50 Gy (^{60}Co gamma-ray), and wound healing was measured. These authors found that autologous adipocyte-derived MSCs improved cutaneous radiation syndrome wound healing, whereas allogeneic adipocyte-derived stem cells did not. In small animals, MSCs collected from different individuals seem not to be an issue [93]. It is evident that MSCs prevent the rejection of allogeneic skin grafts [94].

3.8 Mesenchymal stem cells improve the radiation-induced cognitive dysfunction

Radiotherapy frequently leads to progressive and long-lasting declines in cognition that can severely impact quality of life [95,96,97]. It has been reported recently that MSCs administration restores neuronal plasticity after irradiation. Ref. [36] reported that radiation on brain resulted in cognate dysfunction. When the brain of immunodeficient male athymic nude rats was irradiated at 10 Gy at 2.07 Gy/min, and then total 4×10^5 human neural stem cells (hNSCs) were injected into four different sites of hippocampus 1 month after irradiation, the hNSC transplantation promoted the long-term recovery of host hippocampal neurons and ameliorated cognitive function. The results are stunning and provide insights to further advance research in neuronal injury because of irradiation.

3.9 Mesenchymal stem cells mitigate spinal cord injury

The incidence of spinal cord injury (SCI) is approximately 40.1 cases per million in the United States every year, and males are more likely experience SCI than females [98]. It has been demonstrated that stem cell transplantation therapy, especially with bone marrow mesenchymal stem cells (BMSCs) treatment, mitigates SCI by secreting trophic factors, promoting axonal regeneration and improving neuron survival [99,100]. In addition, after SCI, the BMSCs exhibit effects on amelioration of immunomodulation and glial scarring [101,102]. In most recent publication. When BMSCs at $1 - 3 \times 10^6$ cells in 10 μL volume or mitochondria extracts from BMSCs (3×10^6 cells) were injected into the epicenter of the injured spinal cord of Spague-Dawley male rats. Motor neurons were found to internalize mitochondria and intracellular ATP contents were significantly increased compared to the ATP contents in vehicle-treated neurons. The increases in ATP in neurons between BMSCs-treated and mitochondria-treated SCI rats were similar, suggesting the same protective efficiency offered by both BMSCs and mitochondria.

3.10 Mesenchymal stem cells improve survival after ionizing radiation combined injury

Exposure to ionizing radiation alone or radiation combined with traumatic tissue injury (i.e., radiation combined injury (CI)) is a crucial life-threatening factor in nuclear and radiological accidents. Radiation injuries occur at the molecular, cellular, tissue, and system levels. In our laboratory, we found that B6D2F1/J female mice exposed to ^{60}Co-γ-photon radiation (9.5 Gy, 0.4 Gy/min, bilateral) followed by 15% total-body-surface-area skin wounds (R−W CI) or burns (R−B CI) experienced a higher mortality over a 30-day observation period compared to irradiation alone [75,89]. Radiation CI was accompanied by severe leukocytopenia, thrombopenia, erythropenia, and anemia [91].

This laboratory was the first one to investigate whether treatment with MSCs could improve survival after radiation CI. Bone marrow MSCs were isolated from femurs of B6D2F1/J female mice and were expanded and cultivated in hypoxic conditions (5% O_2, 10% CO_2, 85% N_2) over 30 days. MSCs ($2 - 3 \times 10^6$ cells/mouse) were transfused to mice 24 h after CI because of ^{60}Co-γ-photon irradiation (9.25 and 9.75 Gy, 0.4 Gy/min, bilateral) followed by skin wounding (i.e., radiation CI). Water consumption, body weight, wound healing, and survival tallies were monitored during the observation period. Mice subjected to radiation CI experienced a dramatic moribundity over a 30-day observation period. Thus, radiation CI (9.25 Gy)-animal group was characterized by 40% mortality rate while radiation CI (9.75 Gy)-animal group had 100% mortality rate. Radiation CI-induced sickness was accompanied by body weight loss, increased water intake, and delayed wound healing. At the 30th day post-injury, BMC depletion still remained in surviving radiation CI mice. Treatment of radiation CI (9.25 Gy)-animal group with MSCs led to an increase in 30-day survival rate by 30%, attenuated body weight loss, accelerated wound healing rate, and ameliorated bone-marrow cell depletion. Treatment of radiation CI (9.75 Gy)-animal group with MSCs led to an increase in 30-day survival rate by 20% [11], suggesting that MSC therapy is efficacious to sustain animal survival after radiation CI. It is worthy to note that transfusion of MSCs with 1×10^6 cells failed to improve the survival after radiation CI. Transfusion of MSCs with more than 3×10^6 cells resulted in thrombosis and subsequent lethality in these mice. Thus, possibilities of multiple injections of MSCs less than 3×10^6 cells/mouse providing better desirable survival outcomes will warrant further investigation in this regard.

3.11 Mesenchymal stem cells effectively treat articular cartilage defects and osteoarthritis

Sports injuries, accidental trauma, aging, osteoarthritis, necrosis of subchondral bone, or arthritis can cause articular cartilage damage. There is not only very limited for repair but also may trigger progressive damage and joint degeneration. Current treatments for this kind of damage are primarily to alleviate symptoms, reduce pain, and control inflammation. But the treatments do not control the progressive degeneration of joint tissues [103]. Tissue engineering-based cartilage repair is a major research and was first reported by Ref. [104]; but the result was associated with undesirable disadvantages such as presence of fibrocartilage instead of hyaline cartilage. Therefore, MSCs are targeted. MSCs have an immunomodulatory property, can attenuate tissue injury, inhibit fibrotic remodeling and apoptosis, promote angiogenesis, stimulate stem cell recruitment and proliferation, and reduce oxidative stress [25]. The mediators important during tissue remodeling are MMPs and tissue inhibitors of MMPs (TIMPs). MSCs secret TIMPs that protect implanted cells via TIMP-mediated inhibition of MMP activity [105].

The chondrogenic potential of MSCs was first reported by Ref. [106]. MSCs were delivered to a defective site by direct intra-articular injection or via a scaffold. There were MSC studies in animals [25]. Currently, 20 clinical trials are out there (see www.clinicaltrials.gov). To date, the literature contains information on 15 clinical trials or case reports with follow-up durations of at least 6 months, including pain measurements, assessment of range of motion, and MRI after patients receiving a single injection of MSCs at doses of 50×10^6 cells or 150×10^6 cells. Allogenic and autologous MSCs are similarly effective [107]. There is one clinical result reporting that the single MSC injection did not cause any problem up to two years later and no abnormal tissue growth was apparent [108].

3.12 Mesenchymal stem cells attenuate the severity of acute graft-versus-host disease

Steroid and other immunosuppressants are standard treatments of graft-versus-host disease (GvHD). However, the possibility of lethal infections may still increase and no effective treatments are available for severe steroid-refractory GvHD [109]. MSCs were targeted for treating the problem because of its possible immunosuppressive property. The latter has been controversial. Ref. [110] reported in ex vivo human peripheral blood mononuclear cells (PBMCs) the effector function of virus-specific T cells may be retained after MSC infusion. Ref. [111] also reported that in vivo infusion of $0.9-1.3 \times 10^6$/kg MSCs did not make difference in chronic GvHD development and infectious complications, although the mortality and occurrence of GvHD were reduced in MSC-infused patients. Ref. [21] reported that GvHD patients infused with single or multiple doses of $0.4 - 9 \times 10^6$ cells/kg bodyweight MSCs did not have side-effects during or immediately after infusions of MSCs. Furthermore, MSCs-infused GvHD patients had higher overall survival rate. The data were promising.

In another GvHD-related disease, namely, steroid-resistant acute GvHD (srAGVHD) is the most important cause of morbidity and mortality after allogeneic stem cell transplantation. Ref. [112] reports that MSC application is a treatment method that can be applied safely in combination with other treatment methods in srAGVHD, a condition that has a high-mortality rate. There are almost no acute side effects. There are also no serious long-term side effects in the literature.

Animal models for GvHD have been available for studying the possibility of extending survival. Ref. [113] reported that like human GvHD patients, infusion of 5×10^5 MSCs to GvHD-induced C57BL/6xC3H F1 mice on days 3 and 7 following bone marrow transplantation prevented GvHD induction. To use allogeneic MSCs instead of autologous MSCs, Ref. [35] reported that MSCs were transfected with inducible co-stimulator (ICOS) gene that is a member of CD28 family and essential in T cell activation and differentiation [114]. Then, these allogeneic ICOS-transfected MSCs at 2×10^5 cells/mouse were injected intravenously into GvHD mice. The recipient mice were monitored daily for survival for up to 120 days. The gene transfected MSCs-injected mice survived better and were associated with lower incidence and severity of acute GvHD. These $MSCs^{ICOS}$ were able to suppress Th1 and Th17 polarization and promoted Th2 polarization on both protein expression and gene transcription levels. Higher serum levels of IL-4, IL-10, and lower levels of IFN-γ, IL-2, IL-12, and IL-17A were detected in $MSCs^{ICOS}$-injected mice. The results further reinforce that even allogeneic MSCs injection is also a promising strategy for acute GvHD prevention and treatment.

Animal models for GvHD have been available for elucidating the immunosuppression activity of MSCs. Ref. [113] indicated that the immunosuppressive function of MSCs was elicited by IFNγ and the concurrent comitant presence of any of TNFα, IL-1α, or IL-1β. These combinations provoked the expression of several chemokines and iNOS in MSCs, thereby deriving T cell migration into proximity with MSCs where T cell responsiveness is suppressed by NO. Other laboratories reported that MSCs immunosuppressive function involved with IL-10 [115], TGF-β [116], NO [117], indoleamine 2,3-dioxygenase [118], and proglandin E2 [119]. However, Ref. [120] reported that MSCs are not intrinsically immune-privileged, and under allogenic settings, these cells induce rejection, which is followed by an immune memory. The long-term survival of allogenic MSCs likely represents a major challenge for use of allogenic MSCs for repair or regeneration of tissue and organs.

4. Replacement of mesenchymal stem cells with exosomes including anti-inflammatory cytokines, growth factors, and micro-RNAs

When injecting foreign stem cells into the brain, there remain safety concerns such as immunogenicity and teratoma. When injecting stem cells systemically, lung fibrosis and thrombosis occur. Therefore, it prompts to seek cell-free alternatives. Exosomes receive great attention among cell-free alternatives. Exosomes are nanometer sized particles released from cells in every tissue type. They have lipid membrane, containing extracellular vesicles with diameter of less than 150 nm. Exosomes contain proteins, nucleic acids, lipids, cytokines, microRNAs, etc. They are important for cell-to-cell communications in normal cells and pathological cells [39].

4.1 Exosome characterization

Exosomes can be isolated with three common techniques: a stepwise ultracentrifugation at 100,000xg, an ultrafiltration with 3 kDa exclusion membranes, and a charge-based precipitation method. The diameters are 116.2 nm (ultracentrifugation), 178.7 nm (ultrafiltration), and 453.1 nm (precipitation); the counts of particles/mL are 9.6×10^8 (ultracentrifugation), 52.5×10^9 (ultrafiltration), and 2.02×10^9 (precipitation). The ultrafiltration method gives the best yield. Relevant markers for exosomes are tetraspanins CD9, CD63, and CD81 [120a]. Exosomes can pass blood brain barrier due to their small size.

Ionizing radiation to mouse head induced increases in exosome secretion in a time and dose-dependent manner [121,122]. The difference between MSCs and exosomes is the latter can pass blood brain barrier while the former cannot.

4.2 Exosome therapy

Analysis of metabolic and lipidomic profiling in plasma suggests that relevant markers present in exosomes and plasma can be used for assessing ionizing radiation toxicities to brain [38]. MSCs-derived exosomes can be effective in treating brain injury including radiation-induced cognitive deficits as well [39].

Ref. [37] reported that human neural stem cells (hNSCs, 1×10^5 cells in 2 μL) into four brain sites or exosomes (1×10^8 cells in 2 μL) into two brain sites of male athymic nude rats were injected after head irradiation at 10 Gy (1 Gy/min). One month later, the morphometric improvements were observed with increasing glia cell-derived growth factor and activated microglia and decreasing postsynaptic density protein 95. The data suggest that either hNSCs or exosomes are potential countermeasures against the adverse effects of brain irradiation.

Transplantation of embryonic stem cell—derived NSCs has been thought as a novel potential therapeutic approach for neurodegenerative disorders and psychiatric disorders [123]. MiR-124 is one of the most abundant and the best characterized miRNA specifically expressed in the adult brain [124]. MiR-124 expression was discovered to be overexpressed in neuronal differentiation, both prenatal and postnatal [125]. Ref. [69] reported that miR-124 promoted proliferation and differentiation of NSCs through inactivating Notch pathway by knocking out Notch ligands Delta-like (DLL4).

Another approach is to co-deliver miRNA and MSCs. Ref. [13] reported that independently co-delivered miR-148b and miR-21 mimicked plasmonic nanoparticle conjugates to induce osteogenic differentiation of human adipose stem cells (hASCs). The results demonstrate that sequential miRNA mimic activation results in upregulation of osteogenic markers and mineralization relative to miR-148b alone, or coactivation of miR-148b and miR-21 at the same time.

Radiation, surgery, physical trauma, burn, and diabetes-associated diseases cause skin injury. This wound healing process is complicated because of involvement of multiple organs/systems. Many efficacious candidates and MSCs have been successfully tested out [125a]. Recently, exosomes have been explored. Ref. [17] reported that exosomes from the supernatant of cultured adipose-derived stem cells were isolated using ultracentrifugation and filtration. Then, the exosomes were loaded in the alginate-based hydrogel and applied on the wound site in an animal model. This wound dressing technique has highly improved wound closure, collage synthesis, and vessel formation in the wound area. Ref. [126] also reported that the same adipose stem cell-derived exosomes demonstrated a neuroprotection against amyotrophic lateral sclerosis. These exosomes were internalized by neurons. Antiapoptotic effect of exosomes in these neurons were observed as decreases in Bax protein and active caspase-3 protein and increases in Bcl-2α.

Collectively speaking, cell-free based treatments have more advantages than MSC treatments because MSCs tend to deposit in the lung and kidney, thereby leading to organ fibrosis that impedes the organ function. Exosomes derived from MSCs fulfill not only the need but also exhibit the similar efficacy as MSCs do. However, synthetic/engineered exosomes that contain miR-124, IL-10, growth factors, and other prolific molecules may be even more desirable to meet needs of tissue repair and therapy.

5. Conclusions

MSCs have been extensively characterized. Promising experimental and clinical data are beginning to emerge to support the use of MSCs for repair or regeneration of tissues and organs. Therefore, MSCs as an effective therapy for patients/victims are promising. MSCs can be easily harvested from bone marrow and fat tissues. They can be cultured, grown, and expanded in the laboratory for mass production. Therefore, meeting commercial needs for health maintenance or tissue repair and regeneration can be envisioned and accomplished. With advantage of autologous transplantation which avoids the immune rejection and ethical concerns, MSCs have great application prospect in personalized treatment of diseases. However, the new technology advance allows to isolate exosomes (~150 nm in diameter) that induce similar efficacy as MSC treatment, without known side effects from MSCs.

Acknowledgments

The author thanks AFRRI management and leadership and the USUHS External Affairs Office for clearing/approving this manuscript for publication. The views, opinions, and findings contained in this report are those of the author and do not reflect official policy or positions of the Armed Forces Radiobiology Research Institute, the Uniformed Services University of the Health Sciences, the Department of Defense, The National Institutes of Health, or the United Stated government. The commercial products identified in this document do not imply recommendation or endorsement by the Federal Government and do not imply that the products identified herein are necessarily the best available for the purpose. Research was supported by NIAID YI-AI-5045-04 and AI080553, AFRRI RAB33336, RAB33529, RBB34363 and RAB310934, and JPC-7 VP000276-01 (to JGK).

References

[1] Hu Z-B, Zeng R, Guo W-T, Lin H. Induction and differentiation of bone marrow mesenchymal stem cells. J Clin Rehabilitative Tissue Eng Res 2008;12(43):8561–6.

[2] Kiang JG. Adult mesenchymal cells and radiation injury. Health Phys 2016;111(2): 198–203.

[3] Kolf CM, Cho E, Tuan RS. Mesenchymal stromal cells. Biology of adult mesenchymal stem cells: regulation of niche, self-renewal and differentiation. Arthritis Res Ther 2007;9(1):204.

[4] Erices A, Conget P, Minguell JJ. Mesenchymal progenitor cells in human umbilical cord blood. Br J Haematol 2000;109(1):235–42.

[5] Ikarashi S, Tsuchiya A, Kawata Y, Kojima Y, Watanabe T, Takeuchi S, et al. Effects of human adipose tissue-derived and umbilical cord tissue-derived mesenchymal stem cells in a dextran sulfate sodium-induced mouse model. Biores Open Access 2019; 8(1):185–99.

[6] Lee OK, Kuo TK, Chen W-M, et al. Isolation of multipotent mesenchymal stem cells from umbilical cord blood. Blood 2004;103(5):1669–75.

[7] Schmelzer E, McKeel DT, Gerlach JC. Characterization of human mesenchymal stem cells from different tissues and their membrane encasement for prospective transplantation therapies. Biomed Res Int 2019;2019:6376271.

[8] Kern S, Eichler H, Stoeve J, et al. Comparative analysis of mesenchymal stem cells from bone marrow, umbilical cord blood, or adipose tissue. Stem Cells 2006;24(5): 1294–301.

[9] Yen BL, Huang H-I, Chien C-C, et al. Isolation of multipotent cells from human term placenta. Stem Cells 2005;23(1):3–9.

[10] In't Anker PS, Scherjon SA, Kleijburg-Van Der Keur C, et al. Isolation of mesenchymal stem cells of fetal or maternal origin from human placenta. Stem Cells 2004;22(7):1338–45.

[11] Kiang JG, Gorbunov NV. Bone marrow mesenchymal stem cells increase survival after ionizing irradiation combined with wound trauma: characterization and therapy. J Cell Sci Ther 2014;5:6.

[12] Tropel P, Noel D, Platet N, et al. Isolation and characterisation of mesenchymal stem cells from adult mouse bone marrow. Exp Cell Res (ECR) 2004;295(2):395–406.

[13] Abu-Laban M, Hamal P, Arrizabalaga JH, Forghani A, Dikkumbura AS, Kumal RR, et al. Combinatorial delivery of miRNA-nanoparticle conjugates in human adipose stem cells for amplified osteogenesis. Small 2019;14:e1902864.

[14] Gimble JM, Katz AJ, Bunnell BA. Adipose-derived stem cells for regenerative medicine. Circ Res 2007;100(9):1249–60.

[15] Krishnan B, Sallam HS, Tumurbataar B, Saieva S, Baymon D, Tuvdendorj D, et al. Amelioration of hippocampal dysfunction by adipose tissue-targeted stem cell transplantation in a mouse model of type 2 diabetes. J Neurochem 2020;153(1):51–62. https://doi.org/10.1111/jnc.14915.

[16] Lee J, Kim OH, Lee SC, Kim KH, Shin JS, Hong HE, et al. Enhanced therapeutic potential of the secretome released from adipose-derived stem cells by PGC-1α-driven upregulation of mitochondrial proliferation. Int J Mol Sci 2019;20(22). E5589.

[17] Shafei S, Khanmohammadi M, Heidari R, Ghanbari H, Taghdiri Nooshabadi V, Farzamfar S, et al. Exosome loaded alginate hydrogel promotes tissue regeneration in full-thickness skin wounds: an in-vivo study. J Biomed Mater Res A 2020; 108(3):545–56. https://doi.org/10.1002/jbm.a.36835.

[18] Sikora B, Skubis-Sikora A, Kimsa-Furdzik M, Ciszek W, Kostrzewski M, Stojko J, et al. Adipose-derived stem cells undergo differentiation after co-culture with porcine limbal epithelial stem cells. Stem Cell Res 2019;41:101609.

[19] Chamberlain G, Fox J, Ashton B, et al. Concise review: mesenchymal stem cells: their phenotype, differentiation capacity, immunological features, and potential for homing. Stem Cells 2007;25(11):2739–49.
[20] Djouad F, Bouffi C, Ghannam S, et al. Mesenchymal stem cells: innovative therapeutic tools for rheumatic diseases. Nat Rev Rheumatol 2009;5(7):392–9.
[21] Le Blanc K, Frassoni F, Ball L, et al. Mesenchymal stem cells for treatment of steroid-resistant, severe, acute graft-versus-host disease: a phase II study. Lancet 2008; 371(9624):1579–86.
[22] Tian Y, Deng YB, Huang YJ, et al. Bone marrow-derived mesenchymal stem cells decrease acute graft-versus-host disease after allogeneic hematopoietic stem cells transplantation. Immunol Invest 2008;37(1):29–42.
[23] Shen H, Wang Y, Zhang Z, et al. Mesenchymal stem cells for cardiac regenerative therapy: optimization of cell differentiation strategy. Stem Cell Int 2015;2015: 524756.
[24] Strauer BE, Brehm M, Zeus T, et al. Intracoronary, human autologous stem cell transplantation for myocardial regeneration following myocardial infarction. Dtsch Med Wochenschr (DMW) 2001;126(34–35):932–8.
[25] Wang Y, Yuan M, Guo QY, et al. Mesenchymal stem cells for treating articular cartilage defects and osteoarthritis. Cell Transplant 2015;24(9):1661–78.
[26] Fukushima S, Coppen SR, Lee J, et al. Choice of cell-delivery route for skeletal myoblast transplantation for treating postinfarction chronic heart failure in rat. PLoS One 2008;3(8):e3071.
[27] Ghostine S, Carrion C, Souza LCG, et al. Long-term efficacy of myoblast transplantation on regional structure and function after myocardial infarction. Circulation 2002;106(12 Suppl. 1):I131–6.
[28] Hao L, Wang J, Zou Z, et al. Transplantation of BMSCs expressing hPDGF-A/hBD2 promotes wound healing in rats with combined radiation-wound injury. Gene Ther 2009;16(1):34–42.
[29] Francois S, Mouiseddine M, Allenet-Lepage B, et al. Human mesenchymal stem cells provide protection against radiation-induced liver injury by antioxidative process, vasculature protection, hepatocyte differentiation, and trophic effects. Biomed Res Int 2013;2013:151679.
[30] Hu KX, Sun QY, Guo M, et al. The radiation protection and therapy effects of mesenchymal stem cells in mice with acute radiation injury. Br J Radiol 2010; 83(985):52–8.
[31] Saha S, Bhanja P, Kabarriti R, et al. Bone marrow stromal cell transplantation mitigates radiation-induced gastrointestinal syndrome in mice. PLoS One 2011;6(9): e24072.
[32] Garg S, Wang W, Prabath BG, et al. Bone marrow transplantation helps restore the intestinal mucosal barrier after total body irradiation in mice. Radiat Res 2014;181(3): 229–39.
[33] He H, Liu L, Chen Q, et al. Mesenchymal stem cells overexpressing angiotensin-converting enzyme 2 rescue lipopolysaccharide-induced lung injury. Cell Transplant 2015;24(9):1699–715.
[34] Franchi F, Peterson KM, Xu R, et al. Mesenchymal stromal cells improve renovascular function in polycystic kidney disease. Cell Transplant 2015;24(9):1687–98.
[35] Yang D, Wang LP, Zhou H, et al. Inducible costimulator gene-transduced bone marrow-derived mesenchymal stem cells attenuate the severity of acute graft-versus-host disease in mouse models. Cell Transplant 2015;24(9):1717–31.
[36] Acharya MM, Rosi S, Jopson T, et al. Human neural stem cell transplantation provides long-term restoration of neuronal plasticity in the irradiated hippocampus. Cell Transplant 2015;24(4):691–702.

[37] Smith SM, Giedzinski E, Angulo MC, Lui T, Lu C, Park AL, et al. Functional equivalence of stem cell and stem cell-derived extracellular vesicle transplantation to repair the irradiated brain. Stem Cells Transl Med 2020;9(1):93–105. https://doi.org/10.1002/sctm.18-0227.
[38] Hinzman CP, Baulch JE, Mehta KY, Girgis M, Bansal S, Gill K, et al. Plasma-derived extracellular vesicles yield predictive markers of cranial irradiation exposure in mice. Sci Rep 2019;9(1):9460.
[39] Leavitt RJ, Limoli CL, Baulch JE. miRNA-based therapeutic potential of stem cell-derived extracellular vesicles: a safe cell-free treatment to ameliorate radiation-induced brain injury. Int J Radiat Biol 2019;95(4):427–35.
[40] Abdel-Mageed AS, Senagore AJ, Pietryga DW, et al. Intravenous administration of mesenchymal stem cells genetically modified with extracellular superoxide dismutase improves survival in irradiated mice. Blood 2009;113(5):1201–3.
[41] Dayoub H, Dumont RJ, Li JZ, et al. Human mesenchymal stem cells transduced with recombinant bone morphogenetic protein-9 adenovirus promote osteogenesis in rodents. Tissue Eng 2003;9(2):347–56.
[42] Dominici M, le Blanc K, Mueller I, et al. Minimal criteria for defining multipotent mesenchymal stromal cells. The International Society for Cellular Therapy position statement. Cytotherapy 2006;8(4):315–7.
[43] Jackson KA, Majka SM, Wang H, et al. Regeneration of ischemic cardiac muscle and vascular endothelium by adult stem cells. J Clin Invest 2001;107(11):1395–402.
[44] Jorgensen C, Djouad F, Apparailly F, et al. Engineering mesenchymal stem cells for immunotherapy. Gene Ther 2003;10(10):928–31.
[45] Kocher AA, Schuster MD, Szabolcs MJ, et al. Neovascularization of ischemic myocardium by human bone-marrow-derived angioblasts prevents cardiomyocyte apoptosis, reduces remodeling and improves cardiac function. Nat Med 2001;7(4):430–6.
[46] Kraitchman DL, Heldman AW, Atalar E, et al. In vivo magnetic resonance imaging of mesenchymal stem cells in myocardial infarction. Circulation 2003;107(18):2290–3.
[47] Tocci A, Forte L. Mesenchymal stem cell: use and perspective. Hematol J 2003;4(2):92–6.
[48] Tuan RS, Boland G, Tuli R. Adult mesenchymal stem cells and cell-based tissue engineering. Arthritis Res Ther 2003;5(1):32–45.
[49] Geng Y-J. Molecular mechanisms for cardiovascular stemcell apoptosis and growth in the hearts with atherosclerotic coronary disease and ischemic heart failure. Ann N Y Acad Sci 2003;1010:687–97.
[50] Gronthos S, Zannettino ACW, Hay SJ, et al. Molecular and cellular characterization of highly purified stromal stem cells derived from human bone marrow. J Cell Sci 2003;116(9):1827–35.
[51] Lodie TA, Blickarz CE, Devarakonda TJ, et al. Systematic analysis of reportedly distinct populations of multipotent bone marrow-derived stem cells reveals a lack of distinction. Tissue Eng 2002;8(5):739–51.
[52] Friedenstein AJ, Chailakhjan RK, Lalykina KS. The development of fibroblast colonies in monolayer cultures of guinea-pig bone marrow and spleen cells. Cell Tissue Kinet 1970;3(4):393–403.
[53] Pittenger MF, Mackay AM, Beck SC, et al. Multilineage potential of adult human mesenchymal stem cells. Science 1999;284(5411):143–7.
[54] Simmons PJ, Torok-Storb B. Identification of stromal cell precursors in human bone marrow by a novel monoclonal antibody, STRO-1. Blood 1991;78(1):55–62.
[55] Felka T, Ulrich C, Rolauffs B, et al. Nitric Oxide activates signaling by c-Raf, MEK, p-JNK, p38 MAPK and p53 in human mesenchymal stromal cells and inhibits their osteogenic differentiation by blocking expression of Runx2. J Stem Cell Res Ther 2014;4:195.

[56] Moreira A, Kahlenberg S, Hornsby P. Therapeutic potential of mesenchymal stem cells for diabetes. J Mol Endocrinol 2017;59(3):R109–20.
[57] Izadpanah R, Trygg C, Patel B, et al. Biologic properties of mesenchymal stem cells derived from bone marrow and adipose tissue. J Cell Biochem 2006;99(5):1285–97.
[58] Boyer LA, Plath K, Zeitlinger J, et al. Polycomb complexes repress developmental regulators in murine embryonic stem cells. Nature 2006;441(7091):349–53.
[59] Jiang Y, Vaessen B, Lenvik T, et al. Multipotent progenitor cells can be isolated from postnatal murine bone marrow, muscle, and brain. Exp Hematol 2002;30(8):896–904.
[60] Metcalf D. The unsolved enigmas of leukemia inhibitory factor. Stem Cells 2003;21(1):5–14.
[61] Tsutsumi S, Shimazu A, Miyazaki K, et al. Retention of multilineage differentiation potential of mesenchymal cells during proliferation in response to FGF. Biochem Biophys Res Commun 2001;288(2):413–9.
[62] Zaragosi LE, Ailhaud G, Dani C. Autocrine fibroblast growth factor 2 signaling is critical for self-renewal of human multipotent adipose-derived stem cells. Stem Cells 2006;24(11):412–9.
[63] Kléber M, Sommer L. Wnt signaling and the regulation of stem cell function. Curr Opin Cell Biol 2004;16(6):681–7.
[64] Bienz M. The subcellular destinations of APC proteins. Nat Rev Mol Cell Biol 2002;3(5):328–38.
[65] Chen D, Zhao M, Mundy GR. Bone morphogenetic proteins. Growth Factors 2004;22(4):233–41.
[66] Hartmann CA. Wnt canon orchestrating osteoblastogenesis. Trends Cell Biol 2006;16(3):151–8.
[67] Nuttall ME, Gimble JM. Controlling the balance between osteoblastogenesis and adipogenesis and the consequent therapeutic implications. Curr Opin Pharmacol 2004;4(3):290–4.
[68] Dezawa M, Ishikawa H, Itokazu Y, et al. Bone marrow stromal cells generate muscle cells and repair muscle degeneration. Science 2005;309(5732):314–7.
[69] Jiao S, Liu Y, Yao Y, Teng J. miR-124 promotes proliferation and differentiation of neuronal stem cells through inactivating Notch pathway. Cell Biosci 2017;7:68.
[70] De Luca A, Verardi R, Neva A, et al. Comparative analysis of mesenchymal stromal cells biological properties. ISRN Stem Cells 2013;2013:674671.
[71] Jiang C, Sun J, Dai Y, et al. HIF-1A and C/EBPs transcriptionally regulate adipogenic differentiation of bone marrow-derived MSCs in hypoxia. Stem Cell Res Ther 2015;6:21.
[72] Gorbunov NV, Elliott TB, McDaniel DP, Lund K, Liao PJ, Zhai M, Kiang JG. Up-regulation of autophagy defense mechanisms in mouse mesenchymal stromal cells in response to ionizing irradiation followed by bacterial challenge. In: Bailly Y, editor. Autophagy. Rijeka, Croatia: InTech Open Access Publisher; 2013, ISBN 980-953-307-971-9. p. 331–50. www.intechweb.org.
[73] Fukumoto R, Cary LH, Gorbunov NV, et al. Ciprofloxacin modulates cytokine/chemokine profile in serum, improves bone marrow repopulation, and limits apoptosis and autophagy in ileum after whole body ionizing irradiation combined with skin-wound trauma. PLoS One 2013;8(3):e58389.
[74] Kiang JG, Smith JT, Anderson MN, et al. Hemorrhage exacerbates radiation effects on survival, leukocytopenia, thrombopenia, erythropenia, bone marrow cell depletion and hematopoiesis, and inflammation-associated microRNAs expression in kidney. PLoS One 2015;10(9):e0139271.

[75] Kiang JG, Jiao W, Cary LH, et al. Wound trauma increases radiation-induced mortality by activation of iNOS pathway and elevation of cytokine concentrations and bacterial infection. Radiat Res 2010;173(3):319–32.

[76] Balduino A, Hurtado SP, Frazão P, et al. Bone marrow subendosteal microenvironment harbours functionally distinct haemosupportive stromal cell populations. Cell Tissue Res 2005;319(2):255–66.

[77] Greenberger JS, Epperly M. Bone marrow-derived stem cells and radiation response. Semin Radiat Oncol 2009;19(2):133–9.

[78] Krebsbach PH, Kuznetsov SA, Bianco P, et al. Bone marrow stromal cells: characterization and clinical application. Crit Rev Oral Biol Med 1999;10(2):165–81.

[79] Owen M, Friedenstein A. Stromal stem cells: marrow-derived osteogenic precursors. Ciba Found Symp 1988;136:42–60.

[80] Gorbunov NV, Garrison BR, McDaniel DP, et al. Adaptive redox response of mesenchymal stromal cells to stimulation with lipopolysaccharide inflammagen: mechanisms of remodeling of tissue barriers in sepsis. Oxid Med Cell Longev 2013;2013:186795.

[81] Gorbunov NV, McDaniel DP, Zhai M, et al. Autophagy and mitochondrial remodelling in mouse mesenchymal stromal cells challenged with Staphylococcus epidermidis. J Cell Mol Med 2015;19(5):1133–50.

[82] Kunwar A, Adhikary B, Jayakumar S, et al. Melanin, a promising radioprotector: mechanisms of actions in a mice model. Toxicol Appl Pharmacol 2012;264(2): 202–11.

[82a] Kiang J, Smith J, Anderson M, et al. A novel therapy, using Ghrelin with pegylated G-CSF, inhibits brain hemorrhage from ionizing radiation or combined radiation injury. Pharm. Pharmacol. Int. J. 2019;7(3):133–45. https://doi.org/10.15406/ppij.2019.07.00243.

[82b] Kiang J, Smith J, Cannon G, et al. Ghrelin, a novel therapy, corrects cytokine and NF-κB-AKT-MAPK network and mitigates intestinal injury induced by combined radiation and skin-wound trauma. Cell Biosci. 2020;10:63. https://doi.org/10.1186/s13578-020-00425-z.

[82c] Path G, Perakakis N, Mantzoros C, et al. Stem cells in the treatment of diabetes mellitus - Focus on mesenchymal stem cells.Metabolism. Metabolism 2019;90:1–15. https://doi.org/10.1016/j.metabol.2018.10.005.

[83] Ware LB, Matthay MA. The acute respiratory distress syndrome. N Engl J Med 2000; 342(18):1334–49.

[84] Anjos-Afonso F, Siapati EK, Bonnet D. In vivo contribution of murine mesenchymal stem cells into multiple cell-types under minimal damage conditions. J Cell Sci 2004; 117(Pt 23):5655–65.

[85] Torres VE, Harris PC. Polycystic kidney diseases: genes, proteins, animal models, disease mechanisms and therapeutics opportunities. J Intern Med 2007;261(1):17–31.

[86] Torres VE, Harris PC. Autosomal dominant polycystic kidney disease: the last 3 years. Kidney Int 2009;76(2):149–68.

[87] Orlic D, Kajstura J, Chimenti S, et al. Mobilized bone marrow cells repair the infarcted heart, improving function and survival. Proc Natl Acad Sci (USA) 2001; 98(18):10344–9.

[88] Jayaraj JS, Janapala RN, Qaseem A, Usman N, Fathima N, Kashif T, et al. Efficacy and safety of stem cell therapy in advanced heart failure patients: a systematic review with a meta-analysis of recent trials between 2017 and 2019. Cureus 2019;11(9): e5585.

[89] Ledney GD, Elliott TB. Combined injury: factors with potential to impact radiation dose assessments. Health Phys 2010;98(2):145–52.

[90] Suzuki F, Loucas BD, Ito I, Asai A, Suzuki S, Kobayashi M. Survival of mice with gastrointestinal acute radiation syndrome through control of bacterial translocation. J Immunol 2018;201:77–86.
[91] Kiang JG, Garrison BR, Burns TM, et al. Wound trauma alters ionizing radiation dose assessment. Cell Biosci 2012;2:20.
[92] Riccobono D, Agay D, Scherthan H, et al. Application of adipocyte-derived stem cells in treatment of cutaneous radiation syndrome. Health Phys 2012;103(2):120–6.
[93] Németh K, Leelahavanichkul A, Yuen PS, et al. Bone marrow stromal cells attenuate sepsis via prostaglandin E(2)-dependent reprogramming of host macrophages to increase their interleukin-10 production. Nat Med 2009;15(1):42–9.
[94] Xu G, Zhang L, Ren G, et al. Immunosuppressive properties of cloned bone marrow mesenchymal stem cells. Cell Res 2007;17:240–8.
[95] Butler JM, Rapp SR, Shaw EG. Managing the cognitive effects of brain tumor radiation therapy. Curr Treat Options Oncol 2006;7(6):517–23.
[96] Meyers CA, Brown PD. Role and relevance of neurocognitive assessment in clinical trials of patients with CNS tumors. J Clin Oncol 2006;24(8):1305–9.
[97] Abayomi OK. Pathogenesis of irradiation-induced cognitive dysfunction. Acta Oncol 1996;35(6):659–63.
[98] Richards C, MacKenzie N, Roberts S, Escorpizo R. People with spinal cord injury in the United States. Am J Phys Med Rehabil 2017;96:S124–6.
[99] Lin L, Lin H, Bai S, Zheng L, Zhang X. Bone marrow mesenchymal stem cells (BMSCs) improved functional recovery of spinal cord injury partly by promoting axonal regeneration. Neurochem Int 2018;115:80–4.
[100] Zhao H, Cheng L, Du X, Hou Y, Liu Y, Cui Z, et al. Transplantation of cerebral dopamine neurotrophic factor transducted BMSCs in contusion spinal cord injury of rats: promotion of nerve regeneration by alleviating neuroinflammation. Mol Neurobiol 2016;53:187–99.
[101] Ock SA, Baregundi Subbarao R, Lee YM, Lee JH, Jeon RH, Lee SL, et al. Comparison of immunomodulation properties of porcine mesenchymal stromal/stem cells derived from the bone marrow, adipose tissue, and dermal skin tissue. Stem Cell Int 2016;2016:9581350.
[102] Ruzicka J, Machova-Urdzikova L, Gillick J, Amemori T, Romanyuk N, Karova K, et al. A comparative study of three different types of stem cells for treatment of rat spinal cord injury. Cell Transplant 2017;26:585–603.
[103] Schroeppel JP, Crist JD, Anderson HC, et al. Molecular regulation of articular chondrocyte function and its significance in osteoarthritis. Histol Histopathol 2011;26: 377–94.
[104] Brittberg M, Lindahl A, Nilsson A, et al. Treatment of deep cartilage defects in the knee with autologous chondrocyte transplantation. N Engl J Med 1994;331(14): 889–95.
[105] Lozito TP, Tuan RS. Mesenchymal stem cells inhibit both endogenous and exogenous MMPs via secreted TIMPs. J Cell Physiol 2011;226(2):385–96.
[106] Ashton BA, Allen TD, Howlett CR, et al. Formation of bone and cartilage by marrow stromal cells in diffusion chambers in vivo. Clin Orthop Relat Res 1980; 151:294–307.
[107] Tay LX, Ahmad RE, Dashtdar H, et al. Treatment outcomes of alginate-embedded allogenic mesenchymal stem cells versus autologous chondrocytes for the repair of focal articular cartilage defects in a rabbit model. Am J Sports Med 2012;40(1):83–90.
[108] Vangsness CT, Farr J, Boyd J, et al. Adult human mesenchymal stem cells delivered via intra-articular injection to the knee following partial medial meniscectomy: a randomized, double-blind, control study. J Bone Joint Surg Am 2014;96A:90–8.

[109] Salmasian H, Rohanizadegan M, Banihosseini S, et al. Corticosteroid regimens for treatment of acute and chronic graft versus host disease (GvHD) after allogenic stem cell transplantation. Cochrane Database Syst Rev 2010;2010:CD005565.
[110] Karlsson H, Samarasinghe S, Ball LM, et al. Mesenchymal stem cells exert differential effects on alloantigen and virus-specific T-cell responses. Blood 2008;112(3):532−41.
[111] Kuzmina LA, Petinati NA, Parovichnikova EN, et al. Multipotent mesenchymal stromal cells for the prophylaxis of acute graft-versus-host disease-A phase II study. Stem Cell Int 2012;2012:968213.
[112] Bozkurt C, Karaöz E, Adaklı Aksoy B, Aydoğdu S, Fışgın T. The use of allogeneic mesenchymal stem cells in childhood steroid-resistant acute graft-versus-host disease: a retrospective study of a single-center experience. Turk J Haematol 2019;36(3): 186−92.
[113] Ren G, Zhang L, Zhao X, et al. Mesenchymal stem cell-mediated immunosuppression occurs via concerted action of chemokines and nitric oxide. Cell Stem Cell 2008; 2(2):141−50.
[114] Dong C, Juedes AE, Temann UA, et al. ICOS co-stimulatory receptor is essential for T-cell activation and function. Nature 2001;409(6816):97−101.
[115] Batten P, Sarathchandra P, Antoniw JW, et al. Human mesenchymal stem cells induce T cell anergy and downregulate T cell allo-responses via the TH2 pathway: relevance to tissue engineering human heart valves. Tissue Eng 2006;12:2263−73.
[116] Groh ME, Maitra B, Szekely E, et al. Human mesenchymal stem cells require monocyte-mediated activation to suppress alloreactive T cells. Exp Hematol 2005; 33:928−34.
[117] Sato K, Ozaki K, Oh I, et al. Nitric oxide plays a critical role in suppression of T-cell proliferation by mesenchymal stem cells. Blood 2007;109:228−34.
[118] Meisel R, Zibert A, Laryea M, et al. Human bone marrow stromal cells inhibit allogeneic T-cell responses by indoleamine 2,3-dioxygenase-mediated tryptophan degradation. Blood 2004;103:4619−21.
[119] Aggarwal S, Pittenger MF. Human mesenchymal stem cells modulate allogeneic immune cell responses. Blood 2005;105:1815−22.
[120] Zangi L, Margalit R, Reich-Zeliger S, et al. Direct imaging of immune rejection and memory induction by allogeneic mesenchymal stromal cells. Stem Cells 2009;27(11): 2865−74.
[120a] Klymiuk M, Balz N, Elashry M, et al. Exosomes isolation and identification from equine mesenchymal stem cells. BMC Vet. Res. 2019;15(1):42. https://doi.org/10.1186/s12917-019-1789-9.
[121] Cheema AK, et al. Plasma derived exosomal biomarkers of exposure to ionizing radiation in nonhuman primates. Int J Mol Sci 2018;19. https://doi.org/10.3390/ijms19113427.
[122] Jelonek K, Widlak P, Pietrowska M. The influence of ionizing radiation on exosome composition, secretion and intercellular communication. Protein Pept Lett 2016;23: 656−63.
[123] Rolando C, Taylor V. Neural stem cell of the hippocampus: development, physiology regulation, and dysfunction in disease. Curr Top Dev Biol 2014;107:183−206.
[124] Ponomarev ED, Veremeyko T, Barteneva N, Krichevsky AM, Weiner HL. MicroRNA-124 promotes microglia quiescence and suppresses EAE by deactivating macrophages via the C/EBP-alpha-PU.1 pathway. Nat Med 2011;17(1):64−70.
[125] Smirnova L, Grafe A, Seiler A, Schumacher S, Nitsch R, Wulczyn FG. Regulation of miRNA expression during neural cell specification. Eur J Neurosci 2005;21(6): 1469−77.

[125a] Kiang J, Olabisi A. A poly-traumatic hit leading to multi-organ death. Cell Biosci. 2019;9:25. https://doi.org/10.1186/s13578-019-0286-y.
[126] Bonafede R, Brandi J, Manfredi M, Scambi I, Schiaffino L, Merigo F, et al. The anti-apoptotic effect of ASC-exosomes in an in vitro ALS model and their proteomic analysis. Cells 2019;8(9):E1087. pii.

Further reading

[1] Gorbunov NV, Elliott TB, McDaniel DP, et al. Up-regulation of autophagy defense mechanisms in mouse mesenchymal stromal cells in response to ionizing irradiation followed by bacterial challenge. INTECH 2013;2013:331−50.
[2] Li H, Wang C, He T, Zhao T, Chen YY, Shen YL, et al. Mitochondrial transfer from bone marrow mesenchymal stem cells to motor neurons in spinal cord injury rats via gap junction. Theranostics 2019;9(7):2017−35.
[3] Kiang Juliann, Zhai Min, Lin Bin. Co-therapy of Pegylated-G-CSF and Ghrelin for enhancing survival after exposure to lethal radiation. Front Pharmacol 2021;2021: 628018. https://doi.org/10.3389/fphar.2021.628018.
[4] Peng RJ, Jiang B, Ding XP, Huang H, Liao YW, Peng G, et al. Effect of TNF-alpha inhibition on bone marrow-derived mesenchymal stem cells in neurological function recovery after spinal cord injury via the Wnt signaling pathway in a rat model. Cell Physiol Biochem 2017;42:743−52.
[5] Shipounova IN, Zhironkina OA, Bigildeev AE, et al. Proliferative potential of multipotent mesenchymal stromal cells from human bone marrow. Bull Exp Biol Med 2012;152(4):543−7.

Index

Note: Page numbers followed by "f" indicate figures and "t" indicate tables.

A

Acute lung injury (ALI), 97, 167
Age-related macular degeneration (AMD), 123
Alveolar epithelial cells (AECs), 95–96
Appressorium
  induction and differentiation
    glucose-ABL1-TOR signaling, 150
    G-protein/protein kinase C signaling pathway, 148–149
    protein kinase C (PKC), 149–150
    surficial recognition, 148
  maturation and function
    autophagy, 150–151
    glycogen metabolism, 151
    lipid, 151
    melanin, 151
    septin-mediated polarization, 151–152
  penetration and invasive growth
    effectors, 153
    MAPK Mps1 signaling pathway, 153
    MAPK PMK1 signaling pathway, 152–153
Articular cartilage defects, 173
Ataxia telangiectasia mutated (ATM), 62
Autophagy, 150–151, 164–165

B

B cells, 58–60
Bradykinin receptor 2 (B2), 97
Branch retinal vein occlusion (BRVO), 134–136, 135f
Breast cancer, 20–21
Bromodomain containing 7 (BRD7), 63–65

C

Cajal-like cells (CLCs), 2
Calpain2 (Capn2), 10
Cell signaling, 8–9
Cellular prolongations, 7
Ceramide-containing lipid rafts, 57–58
Ceramide synthase (CS), 58–60
Cerebromicrovascular endothelial cells (CMVECs), 62–63
Chemokines, 96
Cholelithiasis, 27
Chronic inflammation, 19
Chronic obstructive pulmonary disease (COPD), 25
Contamination, adverse events by, 105
Corticosteroids (CS), 125–126
Crohn's disease, 19

D

Diabetes, 166
Diapedesis, 130–131
Digestive system, 25–27
Direct mechanisms, radiation-induced vascular effects
  endothelial barrier, 45–55, 53f
    PECAM-1, 53–54
    vascular cell adhesion protein 1 (VCAM1), 54
    vesicular endothelial transcytosis transport, 51–52
    in vitro signal transduction, 49–55
    in vivo models, 46–48
  endothelial cell senescence, 55–67
    ataxia telangiectasia mutated (ATM) protein, 62
    B cells, 58–60
    ceramide synthase (CS), 58–60
    human lung microvascular endothelial cells (HLMVEC), 63–65
    insulin-like growth factor-1 (IGF-1), 65
    Knockout mice deficient, 60
    mammalian target of rapamycin (mTOR), 65
    programmed cell death, 55–61
    pulmonaryartery endothelial cells (PAECs), 60–61

Direct mechanisms, radiation-induced vascular effects (*Continued*)
radiation-induced cell cycle arrest, 66

E

Electron microscopy, 5–7
Endogenous repair systems, 97–98
Enzymatic antioxidants, 128–129
Epithelial cells, 20
Exosomes, 99–100, 175–177
characterization, 175
therapy, 176–177
Extracellular vesicles, 99–100

F

Fallopian tubes, 21–23
Fibroblasts, 11

G

Glial fibrillary acidic protein (GFAP), 121
Glucose-ABL1-TOR signaling, 150
Glucose regulated protein 78,000 (GRP78), 60–61
Glycogen metabolism, 151
G-protein/protein kinase C signaling pathway, 148–149
Graft-*versus*-host disease (GvHD), 173–175

H

Heart disease, 23–24
Heterocellular junctions, 11
Human lung microvascular endothelial cells (HLMVEC), 63–65
Human umbilical vein endothelial cells (HUVECs), 52, 62–63
Hypoxia, 163

I

Immune-mediated pathogenesis, 96
Immunohistochemistry, 7
Inner blood-retinal barrier
astrocytes, 121–122, 121f
endothelial cells, eye vessel walls, 117–120, 118f
fluorescein angiography, 134–136
Müller cells, 120–121
optical coherence tomography (OCT), 134–136
pericytes, 120
retinal blood vessels, 117
retinal nerve fiber layer (RNFL), 117
Insulin-like growth factor-1 (IGF-1), 65
Insulin-like growth factor receptor (IGF1R), 65
Interstitial cells of Cajal (ICC), 2
Intestinal mucosal barrier, 14–17, 15f–16f
In vitro signal transduction, 49–55
In vivo models, 46–48

K

Knockout mice deficient, 60

L

Lipid, 151
Liver fibrosis, 28
Localization methods, 3–7, 4t–5t

M

Mammalian target of rapamycin (mTOR), 65
MAPK Mps1 signaling pathway, 153
MAPK PMK1 signaling pathway, 152–153
Mast cell activation, 71–73
Matrix metalloproteinases (MMPs), 121
Melanin, 151
Mesenchymal stem cells (MSCs)
acute lung injury (ALI), 97
adverse events by contamination, 105
autophagy, 164–165
bradykinin receptor 2 (B2), 97
characterization, 160–166
COVID-19, 100–103
endogenous repair systems, 97–98
exosomes, 99–100, 175–177
characterization, 175
therapy, 176–177
extracellular vesicles, 99–100
hypoxia, 163
immune-mediated pathogenesis, 96
microparticles, 99–100
mitochondrial remodeling, 165

negative markers, 160–161
normoxia, 163
organ barrier function, 97
positive markers, 161–162
role of, 99
SARS COV-2, 95–96
self-renewal and maintenance, 162–163
signal transduction, 166
small extracellular vesicles, 105
source of, 103–105
tissue/organ therapy
acute lung injury, 167
articular cartilage defects, 173
diabetes, 166
graft-*versus*-host disease (GvHD), 173–175
intestinal mucosal barrier, radiation-induced damage in, 169
ionizing radiation combined injury, 172
myocardiac performance, 168
osteoarthritis, 173
radiation-induced cognitive dysfunction, 171
radiation-induced delay in wound healing, 170–171
renovascular function, kidney, 167
repair radiation-induced liver injury, 170
spinal cord injury (SCI), 171
tissue reconstitutive mechanisms, 100–103
tissues, 163–164
Microparticles, 99–100
Mitochondrial remodeling, 165
Molecules mammalian target of rapamycin (mTOR), 65
Myocardial infarction (MI), 9–10

N

Negative markers, 160–161
NF-κB essentialmodulator (NEMO), 65–66
Nitric oxide synthase 2 (NOS2), 12
Normoxia, 163

O

Organ barrier function, 97
Osteoarthritis, 173
Outer blood-retinal barrier
fluorescein angiography, 134–136
hepatocyte growth factor (HGF), 122
melanin, 129–130, 130f
optical coherence tomography (OCT), 134–136
reactive oxygen species, 126–129
retinal detachment (RD), 122
retinal organoids, 133–134
retinal pigment epithelium (RPE), 123–126, 125f, 131–133
samples processing, 131–133
three-dimensional (3D) culture conditions, 133–134
tissue engineering, 131–133
uveitis, 130–131
Outer plexiform-layers (OPLs), 117
Oxidative stress, 62

P

PECAM-1, 53–54
Photoreceptor outer segments (POSs), 123
Physiological functions
cell signaling, 8–9
intestinal mucosal barrier, 14–17, 15f–16f
placental barrier, 13–14
skin barrier, 17–19, 18f
tissues
barrier, 12–13
homeostasis, 9–12
regeneration, 9–12
repair, 9–12
Pigment epithelium-derived growth factor (PEDF), 121
Placental barrier, 13–14
Positive markers, 161–162
Processus adhaerens, 8
Progenitor cells, 10–11
Programmed cell death, 55–61
Proliferative diabetic retinopathy (PDR), 134–136, 135f
Protein kinase C (PKC), 149–150
Pulmonary artery endothelial cells (PAECs), 60–61
Pulmonary disease, 24–25

R

Radiation-induced cell cycle arrest, 66
Radiation-induced inflammation, indirect mechanisms, 70–76
Radiation-induced vascular effects
  activated pro-inflammatory endothelial cells, 68–70
  antioxidant response element (ARE), 68–69
  direct mechanisms. *See* Direct mechanisms, radiation-induced vascular effects
  excessive dilation, 43–44
  indirect mechanisms, 70–79
    astrocyte activation by ionizing radiation, 73–76
    iron, endothelial barrier, 76–79
    mast cell activation, 71–73
    radiation-induced inflammation, indirect mechanisms, 70–76
  late effects, 44
Retinal nerve fiber layer (RNFL), 117
Retinal pigment epithelium (RPE)
  blood-brain barrier (BBB), 115–116
  iBRB. *See* Inner blood-retinal barrier
  oBRB. *See* Outer blood-retinal barrier
  outer nuclear layer (ONL), 115
  tight junctions (TJs), 116

S

SARS-COV-2, 95–96
Senescence-associated beta galactosidase (SA-β-gal), 62
Septin-mediated polarization, 151–152
Skin barrier, 17–19, 18f
Small extracellular vesicles, 105
Spinal cord injury (SCI), 171
Stem cells, 10–11
Superoxide dismutase (SOD), 128–129
Surficial recognition, 148

T

Telocytes (TCs)
  Cajal-like cells (CLCs), 2
  cellular prolongations, 7
  definition, 2–3
  discovery, 2–3
  disease, 19–28
    breast cancer, 20–21
    cholelithiasis, 27
    digestive system, 25–27
    fallopian tubes, 21–23
    heart disease, 23–24
    liver fibrosis, 28
    pulmonary disease, 24–25
    ulcerative colitis, 28
    uterus, 21–23
  electron microscopy, 5–7
  future perspectives, 29–30
  history, 2–3
  identification methods, 3–7
  immunohistochemistry, 7
  interstitial cells of Cajal (ICC), 2
  localization methods, 3–7, 4t–5t
  physiological functions, 7–19
  transmission electron microscopy (TEM), 5
  ultrastructure of, 6t
Tissues
  barrier, 12–13
  homeostasis, 9–12
  reconstitutive mechanisms, 100–103
  regeneration, 9–12
  repair, 9–12
Transmission electron microscopy (TEM), 5

U

Ulcerative colitis, 28
Uterus, 21–23

V

Vascular cell adhesion protein 1 (VCAM1), 54
Vascular endothelial growth factor (VEGF), 12, 121
Vascular permeability factor (VPF), 121
Vesicular endothelial transcytosis transport, 51–52

X

X-linked inhibitor of apoptosis (XIAP)-associated factor 1 (XAF1), 63–65

Z

Zonula occludens (ZO), 117–118, 118f

FSC
www.fsc.org
MIX
Papier | Fördert gute Waldnutzung
FSC® C083411

Kolibri 360 GmbH
Ferdinand-Jühlke-Straße 7
99095 Erfurt, Deutschland
produktsicherheit@kolibri360.de